Compression for Multimedia

Providing a thorough theoretical understanding of lossy compression techniques for image, video, speech, and audio compression, this book also covers the key features of each system, as well as practical applications, implementation issues, and design trade-offs. It presents comparisons of multimedia standards in terms of achieving known theoretical limits, whilst common and distinguishing features of the existing standards are explained and related to the background theory. There is detailed coverage of such topics as the H.264 video coding standard, low-complexity code-based vector quantizers, and the Blahut rate-distortion algorithm. Examples based on real multimedia data are also included, together with end-of-chapter problems to test understanding; algorithms that allow the reader to represent speech and audio signals efficiently; and an appendix on the basics of lossless coding. With an excellent balance of theory and practice, this book is ideal for undergraduate and graduate students, and is also a useful reference for practitioners.

Irina Bocharova is an Associate Professor at the Saint Petersburg State University of Information Technologies, Mechanics, and Optics. She has published over 50 technical papers and is the co-inventor of seven US patents in speech, video, and audio coding. Her current research interests include convolutional codes, communication systems, source coding and its applications to speech, audio, and image coding.

Compression for Multimedia

IRINA BOCHAROVA

St Petersburg State University of Information Technologies, Mechanics, and Optics

CAMBRIDGE
UNIVERSITY PRESS

CAMBRIDGE UNIVERSITY PRESS
Cambridge, New York, Melbourne, Madrid, Cape Town, Singapore, São Paulo, Delhi,
Dubai, Tokyo

Cambridge University Press
The Edinburgh Building, Cambridge CB2 8RU, UK

Published in the United States of America by Cambridge University Press, New York

www.cambridge.org
Information on this title: www.cambridge.org/9780521114325

First published 2010

Printed in the United Kingdom at the University Press, Cambridge

A catalogue record for this publication is available from the British Library

ISBN 978-0-521-11432-5 Hardback

Additional resources for this publication at www.cambridge.org/9780521114325

To
my teachers

Contents

Preface

Compression for Multimedia was primarily developed as class notes for my course on techniques for compression of data, speech, music, pictures, and video that I have been teaching for more than 10 years at the University of Aerospace Instrumentation, St Petersburg.

During spring 2005 I worked at Lund University as the Lise Meitner Visiting Professor. I have used part of this time to thoroughly revise and substantially extend my previous notes, resulting in the present version.

I would also like to mention that this task could not have been fulfilled without support. Above all, I am indebted to my colleague and husband Boris Kudryashov. Without our collaboration I would not have reached my view of how various compression techniques could be developed and should be taught. Boris' help in solving many TEX problems was invaluable. Special thanks go to Grigory Tenengolts who supported our research and development of practical methods for multimedia compression. Finally, I am grateful to Rolf Johannesson who proposed me as a Lise Meitner Visiting Professor and, needless to say, to the Engineering faculty of Lund University who made his recommendation become true! Rolf also suggested that I should give an undergraduate course on compression for multimedia at Lund University, develop these notes, and eventually publish them as a book. Thanks!

Lund, April 2009   *Irina Bocharova*

1 Introduction

According to wikipedia[1] *"multimedia is the use of several different media (e.g. text, audio, graphics, animation, video, and interactivity) to convey information."* In a more narrow sense, multimedia is a set of software and hardware means used to create, store, and transmit information presented in various digital formats. Although multimedia data is a general term referring to any type of information, we will focus on multimedia data such as speech, images, audio, and video that are originally analog waveforms.

Due to the dramatic progress in microelectronic technologies during the last decades, TV, photography, sound and video recording, communication systems etc., which came into the world and during at least half of the previous century were developed as analog systems, have been almost completely replaced by digital systems. At the same time numerous digital areas and systems such as video conferencing via the Internet, IP-telephony, multi-user games etc. using digitized speech, images, audio, and video appeared as well. Relatively recently multimedia computer technologies started to penetrate into education, medicine, scientific research, entertainment, advertising, and marketing, as well as into many other universally important areas.

Everything mentioned above motivates a deep study of multimedia compression and intensive research in this area. In order to use analog multimedia signals in digital systems it is necessary to solve two main problems. The first problem, related to using these kinds of signal in digital systems, is how to convert them into digital forms. However, it is not enough simply to digitize them. The number of bits required to store images, audio or video signals converted into digital form is so large that this circumstance limits the efficiency of the corresponding digital systems. Thus, the second problem is to compress multimedia data in order to transmit them faster and to store them more efficiently.

Typically, digitizing *multimedia signals* with a high precision results in large files containing the obtained *multimedia data*. Surely, the exact meaning of the words "large file" or "small file" depends on the level of existing microelectronic technologies. About 20 years ago when multimedia compression was not an everyday attribute of our lives, a floppy disk of size 1.44 Mbytes was a typical storage medium. Data files of size exceeding one diskette were considered as "huge" files at that time. It seemed completely impossible to store, on any kind of existing memory, for example, a digitized

---

[1] http://en.wikipedia.org/wiki/Multimedia

color image of size 1408×1152 pixels. Each pixel of such an image is represented by 3 bytes and thus the image requires 4.9 Mbytes of memory for storing. Transmitting of a color image of size 288×352 through the Plain Old Telephone Service (POTS) networks also looked extremely impractical. Since POTS networks were originally designed for the transmission of analog data they needed a so-called modem to convert the digital data to be transmitted into analog form. It is easy to compute that transmitting $288 \times 352 \times 24 = 2.4$ Mb through the telephone channel using a standard modem with transmitting rate equal to 33.6 kb/s requires approximately 1 min (72 s).

Nowadays a variety of storage devices of large capacity are offered by different companies. Transmitting images through POTS networks also became a thing of the past. New kinds of wideband channel are introduced. In the late nineties Digital Subscriber Line (DSL) modems were developed. They are used to communicate over the same twisted-pair metallic cable as used by telephone networks. Using not a voice channel band but the actual range of frequencies supportable on a twisted-pair circuit, they provide transmitting rates exceeding 1 Mb/s. So, it might seem that compression will not be needed in the future. But this is not the case. Together with increasing storage and channel capacities our requirements of the quality of digital multimedia also increase. First of all, during the last decade typical resolutions of images and video became significantly higher. For example, 2–4 Mpixel digital photocameras are replaced by 8–10 Mpixel cameras. A color picture taken by a 10 Mpixel camera requires 30 Mbytes of memory for storing, i.e. only 66 uncompressed pictures can be stored on a Compact-Flash (CF) memory of rather large size, 2 Gbytes, say. For this reason each photocamera has an embedded image compression algorithm. Moreover, the majority of photocameras do not have a mode which allows us to store the uncompressed image.

Let us consider another example. One second of video film with resolution 480×720 pixels recorded with 30 frames/s requires approximately 31 Mbytes of memory. It means that only 21 s of this film can be recorded on a 650 Mbytes Compact Disc (CD) without compression. However, neither does using 15.9 Gbyte Digital Versatile Disc (DVD) solve the problem of storing video data since only 8.5 minutes of video film with such resolution and frame rate can be recorded on the DVD of such a capacity. As for transmitting high-resolution video through communication channels, it is still an even more complicated problem than storing it. For example, it takes 4 s to transmit a color image of size 480×720 pixels by using the High Data rate DSL (HDSL) modem with rate 2 Mb/s. It means that a film with such a frame size can be transmitted with frame rate equal to 0.25 frame/s only.

The considered examples show that actually it does not matter how much we can increase telecommunication bandwidth or disk storage capacity, there will always remain a need to compress multimedia data in order to transmit them faster and to store them more efficiently.

Multimedia technologies continuously find new applications that create new problems in the multimedia compression field. Recently, new tendencies in multimedia compression have arisen. One of many newly intensively developed areas is Digital Multimedia Broadcasting (DMB), often called "mobile TV." This technology is used in order to send multimedia data to mobile phones and laptops. The world's first DMB service appeared in South Korea in 2005. This line of development requires a significant revision of existing compression algorithms in order to tailor them to the specific needs of broadcasting systems.

The Depth Image Based Rendering (DIBR) technique for three-dimensional television (3D-TV) systems is another quickly developed area that is closely related to multimedia compression. Each 3D image is represented as the corresponding 2D image and an associated per-pixel "depth" information. As a result, the number of bits to represent a 3D image drastically increases. It requires modifications of known 2D compression techniques in order to efficiently compress 3D images and video.

It is, needless to say, about a variety of portable devices such as Personal Digital Assistant (PDA), smartphone, Portable Media Player (iPOD), and many others intended for loading multimedia contents. Each such device requires a new compression algorithm taking into account its specific features.

Multimedia compression systems can be split into two large classes. The first class is *lossless compression systems*. With lossless compression techniques the original file can be recovered exactly, bit by bit after compression and decompression. To do this we usually use well-known methods of discrete source coding such as Huffman coding, arithmetic coding, or coding based on the Ziv–Lempel algorithms.

Lossy compression implies that we remove or reduce the redundancy of the multimedia data at the cost of changing or distorting the original file. These methods exploit the tradeoff of compression versus distortion. Among lossy techniques are compression techniques based on transform coding (coding of the discrete Fourier transform or discrete cosine transform coefficients, wavelet filtering), predictive coding etc.

In this book lossy compression techniques and their applications to image, video, speech, and audio compression are considered. The book provides rather deep knowledge of lossy compression systems. Modern multimedia compression techniques are analyzed and compared in terms of achieving known theoretical limits. Some implementation issues important for the efficient implementation of existing multimedia compression standards are discussed also.

The book is intended for undergraduate students. The required prerequisite is an elementary knowledge of linear systems, Fourier transforms, and signal processing. Some prior knowledge of information theory and random processes would be useful. The book can be also recommended for graduate students with an interest in compression techniques for multimedia.

The book consists of 10 chapters and an Appendix. In Chapters 2 and 3 basic theoretical aspects of source coding with fidelity criteria are given. In particular, in Chapter 3 the notion of the rate-distortion function is introduced. This function is a theoretical limit for achievable performances of multimedia systems. This chapter can be recommended for readers who are doing research in the multimedia compression area. In order to compare the performances of different quantizers some results of the high-resolution quantization theory are given in Section 3.3. This section can be omitted for readers who have no prior knowledge of information theory.

In Chapters 4, 5, and 6, commonly used coding techniques are described. Chapter 4 is devoted to linear predictive coding. It begins in Section 4.1 with descriptions of discrete-time filters by different means which are presented for the sake of completeness and can be omitted by readers familiar with this subject.

Chapters 7, 8, 9, and 10 are devoted to modern standards for speech, image, video, and audio compression, respectively. For readers who are not familiar with lossless coding techniques, the basics of lossless coding are briefly overviewed in the Appendix.

2 Analog to digital conversion

Analog to digital transformation is the first necessary step to load multimedia signals into digital devices. It contains two operations called sampling and quantization. The theoretical background of sampling is given by the famous sampling theorem. The first attempts to formulate and prove the sampling theorem date back to the beginning of the twentieth century. In this chapter we present Shannon's elegant proof of the sampling theorem. Consequences of sampling "too slowly" in the time and frequency domains are discussed. Quantization is the main operation which determines the quality–compression ratio tradeoff in all lossy compression systems. We consider different types of quantizer commonly used in modern multimedia compression systems.

2.1   Analog and digital signals

First, we introduce some definitions.

- A function $f(x)$ is *continuous* at a point $x = a$ if $\lim_{x \to a} f(x) = f(a)$. We say a function is continuous if it is continuous at every point in its domain (the set of its input values).
- We call a set of elements a *discrete* set if it contains a finite or countable number of elements (elements of a countable set can be enumerated).

In the real world *analog signals* are continuous functions of continuous arguments such as time, space, or any other continuous physical variables, although we often use mathematical models with not continuous analog signals such as the saw-tooth signal. We consider mainly time signals which can take on a continuum of values over a defined interval of time. For example, each value can be a real number.

Discrete signals can be discrete over a set of function values and (or) over a set of argument values. In other words, if the analog time signals are sampled, we call this set of numbers which can take on an infinity of values within a certain defined range, a *discrete-time* or *sampled* system. If the sample values are constrained to belong to a discrete set, the system becomes *digital*.

Such signals as images, speech, audio, and video are originally analog signals. In order to convert them into digital signals, we should perform the following two operations:

- First, the signal has to be sampled (the time axis must be discretized or quantized).
- The second operation is to transform the sample values (the obtained list of numbers) in such a manner that each resulting number belongs to a discrete alphabet. We call this operation *quantization*.

We start with a discussion of *sampling* which is a technique of converting an analog signal with a continuous time axis into real values in discrete-time.

Let $x(t)$ be a continuous time function. *Sampling* is taking samples of this function at time instants $t = nT_s$ for all integer values n, where T_s is called *sampling period*. The value $f_s = 1/T_s$ is called *sampling frequency*. Thus, instead of the function $x(t)$ we study the sequence of samples $x(nT_s)$, $n = 0, 1, 2, \ldots$ The first question is: does sampling introduce distortion of the original continuous time function $x(t)$? The second question is: how does the distortion, if any, depend on the value of T_s? The answers are given by the so-called *sampling theorem* (Whittaker 1915; Nyquist 1928; Kotelnikov 1933; Oliver *et al.* 1948), sometimes known as the Nyquist–Shannon–Kotelnikov theorem and also referred to as the Whittaker–Shannon–Kotelnikov sampling theorem.

Before considering the sampling theorem we briefly review the notions which will be used to prove this theorem.

The Fourier transform of the analog signal $x(t)$ is given by the formula

$$X(f) = \int_{-\infty}^{\infty} x(t)e^{-j\omega t}\, dt$$

where $\omega = 2\pi f$ is the *radian frequency* (f is the frequency in Hz). This function is in general complex, with $X(f) = A(f)e^{j\phi(f)}$, where $A(f) = |X(f)|$ is called the *spectrum* of $x(t)$ and $\phi(f)$ is the *phase*. We can also represent $x(t)$ in terms of its Fourier transform via the *inversion formula*

$$x(t) = \int_{-\infty}^{\infty} X(f)e^{j\omega t}\, df.$$

The Fourier transform is closely related to the Laplace transform. For continuous time functions existing only for $t \geq 0$, the Laplace transform is defined as a function of the complex variable s by the following formula

$$L(s) = \int_{0}^{\infty} x(t)e^{-st}\, dt.$$

For $s = j\omega$ the Laplace transform for $x(t)$, $t \geq 0$ coincides with the Fourier transform for this function, if $L(s)$ has no poles on the imaginary axis.

If $x(t)$ is a periodic time function with period p it can be represented as the Fourier series expansion

$$x(t) = \frac{a_0}{2} + \sum_{k=1}^{\infty}(a_k\cos(kt\omega_p) + b_k\sin(kt\omega_p))$$

where

$$\omega_p = \frac{2\pi}{p},$$

$$a_0 = \frac{2}{p} \int_{-p/2}^{p/2} x(t)\,dt,$$

$$a_k = \frac{2}{p} \int_{-p/2}^{p/2} x(t) \cos(kt\omega_p)\,dt, \quad k = 1, 2, \ldots,$$

$$b_k = \frac{2}{p} \int_{-p/2}^{p/2} x(t) \sin(kt\omega_p)\,dt, \quad k = 1, 2, \ldots$$

To replace two of the integrals ($\cos(kt\omega_p)$ and $\sin(kt\omega_p)$) by one for each index k (after simple derivations) we obtain

$$x(t) = \frac{a_0}{2} + \sum_{k=1}^{\infty} A_k \cos(kt\omega_p + \varphi_k)$$

where φ_k is the initial phase.

Using Euler's formula $e^{j\omega t} = \cos(\omega t) + j \sin(\omega t)$ we obtain the complex form of the Fourier series expansion

$$x(t) = \frac{a_0}{2} + \sum_{k=1}^{\infty} \frac{A_k}{2} \left(\exp(jkt\omega_p + j\varphi_k) + \exp(-jkt\omega_p - j\varphi_k) \right)$$

$$= \sum_{k=-\infty}^{\infty} c_k e^{jkt\omega_p}$$

where

$$c_k = \frac{1}{2} A_k e^{j\varphi_k} = \frac{1}{2}(a_k - jb_k) = \frac{1}{p} \int_{-p/2}^{p/2} x(t) \exp(-jkt\omega_p)\,dt,$$

$$c_0 = 1/2 a_0, \, b_0 = 0,$$

$$a_k = a_{-k}, \, b_k = -b_{-k}.$$

It is said that the Fourier series expansion for a periodical function with period p decomposes this function into a sum of harmonical functions with frequencies $k\omega_p$, $k = 1, 2, \ldots$ The Fourier transform for a nonperiodical function represents this function as a sum of an infinite number of harmonical functions with frequencies which differ in infinitesimal quantities. Notice that a nonperiodical function of finite length T also can be decomposed into the Fourier series expansion. To do this we have to construct its periodical continuation with period T.

2.1.1   Sampling theorem

If $x(t)$ is a signal whose Fourier transform $X(f) = \int_{-\infty}^{\infty} x(t)e^{-j2\pi ft}\, dt$ is identically zero $X(f) = 0$ for $|f| > f_{\text{H}}$, then $x(t)$ is completely determined by its samples taken every $1/(2f_{\text{H}})$ s. The frequency $f_{\text{s}} = 1/T_{\text{s}} = 2 f_{\text{H}}$ Hz is called the Nyquist sampling rate.

Proof. Since $X(f) = 0$ for $|f| > f_{\text{H}}$ we can continue $X(f)$ periodically. Then we obtain the periodical function $\hat{X}(f)$ with period equal to $2 f_{\text{H}}$. The function $\hat{X}(f)$ can be decomposed into the Fourier series expansion

$$\hat{X}(f) = \sum_{k=-\infty}^{\infty} a_k e^{j2\pi fk/(2f_{\text{H}})}$$

where

$$a_k = \frac{1}{2f_{\text{H}}} \int_{-f_{\text{H}}}^{f_{\text{H}}} \hat{X}(f)e^{-j2\pi fk/(2f_{\text{H}})}\, df. \qquad (2.1)$$

Since $X(f)$ is the Fourier transform of $x(t)$ then $x(t)$ can be represented in terms of its Fourier transform via the inversion formula

$$x(t) = \int_{-\infty}^{\infty} X(f)e^{j2\pi ft}\, df = \int_{-f_{\text{H}}}^{f_{\text{H}}} \hat{X}(f)e^{j2\pi ft}\, df.$$

Consider the values of the time function $x(t)$ in the discrete points $t = k/(2 f_{\text{H}})$ for all integers k. They can be expressed as follows

$$x\left(\frac{k}{2f_{\text{H}}}\right) = \int_{-f_{\text{H}}}^{f_{\text{H}}} \hat{X}(f)e^{j2\pi fk/(2f_{\text{H}})}\, df. \qquad (2.2)$$

Comparing (2.1) and (2.2), we obtain that

$$a_k = \frac{1}{2f_{\text{H}}} x\left(\frac{-k}{2f_{\text{H}}}\right).$$

Thus, if the time function $x(t)$ is known at points $\ldots, -2/(2f_{\text{H}}), -1/(2f_{\text{H}}), 0, 1/(2f_{\text{H}}), 2/(2f_{\text{H}}), \ldots$ then the coefficients a_k are determined. These coefficients in turn determine $\hat{X}(f)$ and thereby they determine $X(f)$. On the other hand, $X(f)$ determines $x(t)$ for all values of t. It means that there exists a unique time function which does not contain frequencies higher than f_{H} and passes through the given sampling points spaced $1/(2f_{\text{H}})$ s.

In order to reconstruct the time function $x(t)$ using its sampling points $x(k/(2f_{\text{H}}))$ we notice that

$$X(f) = \begin{cases} \sum_{k=-\infty}^{\infty} a_k e^{j2\pi fk/(2f_{\text{H}})}, & \text{if } |f| \le f_{\text{H}} \\ 0, & \text{if } |f| > f_{\text{H}}. \end{cases}$$

To simplify our notations we introduce the *sinc-function* which is defined as

$$\text{sinc}(x) = \frac{\sin(\pi x)}{\pi x}.$$

Using the inverse transform we obtain

$$x(t) = \int_{-f_H}^{f_H} X(f) e^{j2\pi f t}\, df$$

$$= \int_{-f_H}^{f_H} \sum_{k=-\infty}^{\infty} a_k e^{j2\pi f(\frac{k}{2f_H}+t)}\, df$$

$$= \sum_{k} a_k \int_{-f_H}^{f_H} e^{j2\pi f(\frac{k}{2f_H}+t)}\, df$$

$$= \sum_{k=-\infty}^{\infty} a_k \left(\int_{-f_H}^{f_H} \underbrace{\cos\left(2\pi f\left(\frac{k}{2f_H}+t\right)\right)}_{\text{even}}\, df \right.$$

$$\left. + j \int_{-f_H}^{f_H} \underbrace{\sin\left(2\pi f\left(\frac{k}{2f_H}\right)\right)}_{\text{odd}}\, df \right)$$

$$= 2 \sum_{k=-\infty}^{\infty} a_k \int_{0}^{f_H} \cos\left(2\pi f\left(\frac{k}{2f_H}+t\right)\right)\, df$$

$$= 2 f_H \sum_{k=-\infty}^{\infty} a_k \operatorname{sinc}\left(2 f_H\left(\frac{k}{2f_H}+t\right)\right).$$

Simple derivations complete the proof:

$$x(t) = 2 f_H \sum_{k=-\infty}^{\infty} a_k \operatorname{sinc}(2 f_H t + k)$$

$$= \sum_{k=-\infty}^{\infty} x\left(\frac{-k}{2f_H}\right) \operatorname{sinc}(2 f_H t + k)$$

$$= \sum_{i=-\infty}^{\infty} x\left(\frac{i}{2f_H}\right) \operatorname{sinc}(2 f_H t - i),\, i = -k. \qquad (2.3)$$

$\square$

In other words, the time function $x(t)$ can be represented as a sum of elementary functions in the form $\operatorname{sinc}(\alpha)$, $\alpha = 2 f_H t - i$, centered in the sampling points. The sinc-function $\operatorname{sinc}(\alpha)$ is shown in Fig. 2.1. It is equal to 1 in the point $\alpha = 0$, that is, $t = i/(2 f_H)$ and is equal to zero in other sampling points.

It follows from (2.3) that at time instants $t = k T_s = k/(2 f_H)$ the values of $x(t)$ coincide with the sample values $x(k/(2 f_H))$. For the other time instants, it is necessary to sum up an infinite number of series terms in order to reconstruct the exact value of $x(t)$. Therefore, we conclude that in order to reconstruct the function $x(t)$, it is necessary to generate an infinite train of impulses which have form $\operatorname{sinc}(\alpha)$ and are proportional to samples, and to summarize them.

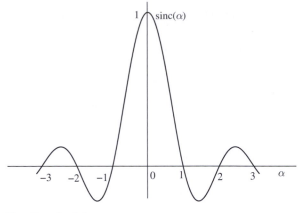

Figure 2.1   Function $\mathrm{sinc}(\alpha)$

The representation of $x(t)$ in the form (2.3) is a particular case of the so-called *orthogonal decomposition* of the function over a system of basis functions. In our case the role of basis functions is played by the sinc-functions $\mathrm{sinc}(2f_{\mathrm{H}}t - i)$ which we call *sampling functions*. They are orthogonal since

$$\int_{-\infty}^{\infty} \mathrm{sinc}(2f_{\mathrm{H}}t - j)\mathrm{sinc}(2f_{\mathrm{H}}t - i)\,dt = \begin{cases} 1/(2f_{\mathrm{H}}), & \text{if } i = j \\ 0, & \text{if } i \neq j. \end{cases}$$

2.1.2   Historical background

The sampling theorem has a rather long history. It started in 1897 when the theorem was partly proved by the French mathematician Emil Borel. He showed that any continuous-time function $x(t)$ whose spectrum is limited by the maximal frequency f_{H} is uniquely defined by its samples with frequency $2f_{\mathrm{H}}$. However, he wrote nothing about how to reconstruct $x(t)$ from these samples. Then in 1915 the English mathematician Edmund Whittaker almost completed the proof by finding the so-called "cardinal function" which had the form

$$\sum_{k} x(kT_{\mathrm{s}})\mathrm{sinc}\left(\frac{t}{T_{\mathrm{s}}} - k\right)$$

but he never stated that this reconstructed function coincides with the original function $x(t)$. Kinnosuke Ogura actually was the first who in 1920 proved the sampling theorem (ignoring some theoretical nuances). In 1928 Harry Nyquist improved the proof by Ogura and in 1933 (independently of Nyquist) Vladimir Kotelnikov published the theorem in its contemporary form. In 1948 Claude Shannon, who was not aware of Kotelnikov's results relying only on Nyquist's proof, formulated and proved the theorem once more.

It follows from the sampling theorem that, if the Fourier transform of a time function is nonzero over a finite frequency band, then taking samples of this time function in the infinite time interval with a sampling rate twice the maximal frequency in the signal spectrum we can perfectly recover the original function from its samples.

2.1.3   Aliasing

The error caused by sampling too slowly (with a sampling rate less than the Nyquist sampling rate) is known as *aliasing*. This name is derived from the fact that the higher frequencies disguise themselves as lower frequencies. In other words, aliasing is an effect that causes different continuous signals to become indistinguishable (or aliases of one another) when sampled. We can observe the same phenomenon when a rotating wheel is viewed as a sequence of individual frames. If the frame rate is not large enough, then the rotating wheel can look as if it is standing still or even rotating in the reverse direction.

The analysis of aliasing is most easily performed in the frequency domain. Before doing that, we will illustrate the problem in the time domain. Figure 2.2 shows a cosinusoidal signal at a frequency of 3 Hz. Assume that this cosinusoid is sampled at the rate of 4 samples/s. According to the sampling theorem the minimum sampling rate which is required to guarantee the unique reconstruction is 6 samples/s, so 4 samples/s is not fast enough. The samples at the rate of 4 samples/s are indicated in Fig. 2.2 by an asterisk. It is easy to see that the same samples would arise from a cosinusoid at 1 Hz, as shown by the dashed curve. The 3 Hz signal is disguising itself as a 1 Hz signal.

Now consider the aliasing effect in the frequency domain. To analyze the Fourier transform of the sampled signal which represents a sequence of numbers, we will associate an analog signal to this sequence. The standard way to associate an analog signal to the sequence of samples is to view the process of sampling as taking the product of the signal $x(t)$ with a train of impulses spaced T_s s apart. To describe the "ideal" impulse,

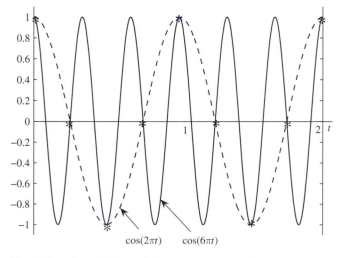

Figure 2.2   Illustration of an aliasing error

we will use the so-called Dirac's delta function $\delta(x)$. It is defined by the following equality:

$$\int_{\Delta} \delta(x)\varphi(x)\,dx = \begin{cases} \varphi(0), & \text{if } 0 \in \Delta \\ 0, & \text{otherwise} \end{cases}$$

where $\varphi(x)$ is an arbitrary continuous function and Δ is an arbitrary interval of the real line. It is evident that

$$\int_{\Delta} \delta(x - x_0)\varphi(x)\,dx = \begin{cases} \varphi(x_0), & \text{if } x_0 \in \Delta \\ 0, & \text{otherwise} \end{cases}$$

and

$$\int_{\Delta} \delta(x - x_0)\,dx = \begin{cases} 1, & \text{if } x_0 \in \Delta \\ 0, & \text{otherwise.} \end{cases}$$

The delta function can be multiplied by any complex number and added to other delta functions or conventional integrable functions. It can also be considered as the limit of the sequence of rectangular impulses of width Δ and height $1/\Delta$ when Δ tends to zero.

If we consider the sequence of samples as the product of the signal $x(t)$ with a train of the ideal impulses spaced T_s s apart, then sampling yields the following signal

$$x_s(t) = \sum_{k=-\infty}^{k=\infty} x(kT_s)\,\delta(t - kT_s). \tag{2.4}$$

Since the function $\delta(t - kT_s)$ is equal to zero everywhere except at the point kT_s we can rewrite (2.4) as

$$x_s(t) = x(t) \sum_{k=-\infty}^{\infty} \delta(t - kT_s). \tag{2.5}$$

Taking into account that $\sum_{k=-\infty}^{\infty} \delta(t - kT_s)$ is a periodical signal with period T_s it can be represented as the Fourier series expansion

$$\sum_{k=-\infty}^{\infty} \delta(t - kT_s) = \sum_{n=-\infty}^{\infty} c_n e^{j2\pi tn/T_s} \tag{2.6}$$

where

$$c_n = \frac{1}{T_s} \int_{-T_s/2}^{T_s/2} \delta(t) e^{-j2\pi tn/T_s}\,dt = \frac{1}{T_s}.$$

We took into account that there is only one delta function inside the interval $(-T_s/2, T_s/2)$ (it corresponds to $k = 0$). Substituting (2.6) into (2.5) we obtain

$$x_s(t) = \frac{1}{T_s} \sum_{n=-\infty}^{\infty} x(t) e^{j2\pi tn/T_s}.$$

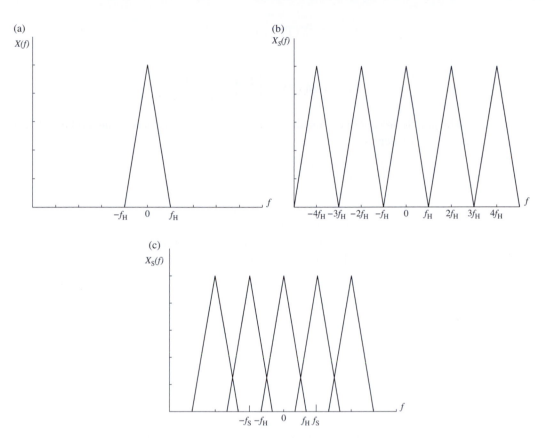

Figure 2.3 Illustration of aliasing error in the frequency domain. (a) Spectrum of $x(t)$. (b) Spectrum of sampled $x(t)$ if $T_s = 1/(2 f_H)$. (c) Spectrum of sampled $x(t)$ if $T_s > 1/(2 f_H)$

Since multiplication of a function by $\exp(j2\pi\alpha t)$ corresponds to a shift of the Fourier transform by α we obtain

$$X_s(f) = \frac{1}{T_s} \sum_{n=-\infty}^{\infty} X\left(f - \frac{n}{T_s}\right).$$

It means that the sampled version of the signal $x(t)$ has a spectrum that can be written as copies of $X(f)$, that is, the spectrum of $x(t)$, spaced $1/T_s$ Hz apart.

Figure 2.3 illustrates the spectrum $X_s(f)$ for the two cases $T_s = 1/(2 f_H)$ and $T_s > 1/(2 f_H)$.

It is easy to see that if the sampling of the original signal is too slow, then the copies of its spectrum overlap which leads to an aliasing error. It follows from Fig. 2.3 that in order to reconstruct the signal $x(t)$ from the sequence of its samples, it is sufficient to filter this sequence by the ideal lowpass filter with cutoff frequency equal to $f_H = 1/(2T_s)$ Hz.

More formally, the Fourier transform of the reconstructed signal $x(t)$ can be expressed as follows

$$X(f) = X_s(f)T_s H_L(f)$$

where

$$H_L(f) = \begin{cases} 1, & \text{if } |f| \le f_H \\ 0, & \text{otherwise.} \end{cases}$$

The impulse response of this filter represents the inverse Fourier transform of $H_L(f)$ and can be written as

$$h_L(t) = \int_{-\infty}^{\infty} H_L(f)e^{j2\pi ft}\, df = 2\int_0^{f_H} \cos(2\pi ft)\, df = \frac{1}{T_s}\text{sinc}\left(\frac{t}{T_s}\right).$$

The sampled signal $x_s(t)$ represents a sum of weighted delta functions. Passing through the ideal lowpass filter each of these delta functions generates a shifted and weighted copy of the filter impulse response. Thus, the output signal represents a sum of weighted and shifted copies of the impulse response of the ideal lowpass filter; that is,

$$x(t) = \sum_{k=-\infty}^{\infty} x(kT_s)\text{sinc}\left(\frac{t - kT_s}{T_s}\right).$$

The sampling theorem requires that the samples are taken at equally spaced time instants on the infinite time axis, and that every sample is used to reconstruct the value of the original function at any particular time. In a real system, the signal is observed during a limited time. The Fourier transform of any time-limited signal is nonzero outside any finite frequency band. How can we apply the sampling theorem to such signals? Usually we restrict our signals to frequencies inside a certain finite band, $[-f_H, f_H]$ say. We choose the value f_H in such a manner that a certain fraction θ (approximately 0.90–0.99) of the energy of the signal lies inside the frequency band $[-f_H, f_H]$. In other words, to find f_H we have to solve the following equation

$$\frac{\int_{-f_H}^{f_H} |X(f)|^2\, df}{\int_{-\infty}^{\infty} |X(f)|^2\, df} = \theta.$$

We call f_H the *effective bandwidth* with respect to the threshold θ.

Sampling of time-limited signals always introduces an aliasing error owing to an overlap of copies of the signal spectrum (see Fig. 2.3(c)).

In practice different techniques are used to reduce the aliasing error. For example, prefiltering the time-limited signal in the analog domain by a proper lowpass filter attenuates all high-frequency spectrum components, reducing the so-called aliasing effect. The other possibility is to choose $T_s \ll 1/(2f_H)$ s. However, since our goal is to represent analog signals in a compact form, we should not sample them too fast. On the other hand, in many signal processing applications we deal only with sampled input signals and actually perform only sampling rate conversions. Thereby, a rather commonly used approach consists of prefiltering of the oversampled input signal by a digital lowpass filter followed by *downsampling*, that is, taking samples with a given step. For example, let the oversampled speech signal be obtained by sampling with the sampling rate 32 kHz. After prefiltering with cutoff frequency equal to 4 kHz, downsampling is performed with step 4 (we take each fourth sample) and the resulting sampling rate is equal to 8 kHz. Notice that the ideal lowpass filter with cutoff frequency $f_s/2$ has an infinite noncausal impulse response in the form of the sinc-function that can be only approximated. For methods applicable to the design of digital filters for sampling rate conversion systems see, for example, Crochiere and Rabiner (1983).

2.2 Scalar quantization

The second of the two operations required to change an analog signal into a digital signal is called *quantization*. We say that quantization is a procedure of transforming sample values into integer values with a finite length of binary representation. The oldest example of quantization is rounding off. Any real number x can be rounded off to the nearest integer, y say, with a resulting quantization error $y - x$. We will start our consideration with *scalar quantization*, that is, the simplest form of the quantization dealing with a single sample value. After this short introduction we give two formal definitions.

- *Scalar quantization* is a mapping Q of a real value x of a continuous random variable X into the closest (in terms of the chosen distortion measure d) value $y = Q(x)$ from a discrete set $Y = \{y_1, y_2, \ldots, y_M\}$. The values y_i, $i = 1, 2, \ldots, M$, are called *output levels*, *reproduction levels*, or are sometimes referred to as *approximating values*. Y is called *approximating set* or *codebook*.
- *Scalar quantizer* is determined by the set of *thresholds* $(t_0, t_1, \ldots, t_M)$, $t_i < t_{i+1}$, $i = 0, 1, \ldots, M - 1$, which split the real line $\mathcal{R}$ into subintervals or *cells* $\Delta_i = (t_{i-1}, t_i]$, which are disjoint, that is, $\Delta_i \bigcap \Delta_j = \emptyset$, and exhaustive, that is, $\bigcup_{i=1}^{M} \Delta_i = \mathcal{R}$, in such a way, that $y_i = Q(x)$ if and only if $x \in \Delta_i$.

It follows from the definition that the thresholds form an increasing sequence and the width of the ith cell is its length $t_i - t_{i-1}$. The reproduction value $y_i \in \Delta_i$. Notice that the two outermost cells, $(t_0, t_1]$ and (t_{M-1}, t_M), are finite if we quantize an interval of $\mathcal{R}$ and infinite if $\mathcal{R}$ is quantized. The first case corresponds to a finite set of cells and M is a finite number. In the second case, a set of cells can be a countable set, that is, M can be infinite. The general definition reduces to the rounding off example if $\Delta_i = (i - 1/2, i + 1/2]$ and $y_i = i$ for all integers i.

The quantization procedure implies that an input value x is sequentially compared with the thresholds t_i, $i = 1, 2, \ldots, M$, in order to specify the cell Δ_i, which contains x. The binary representation of the number i is stored or transmitted over a communication channel.

Quantization always introduces an error called *quantization noise*. The quality of any scalar quantizer is usually measured by a *rate-distortion function* $R(D)$, where

$$D(Q) = \mathrm{E}\{d(x, Q(x))\}$$

denotes the average quantization error, $d(x, Q(x))$ is a non-negative function called *distortion measure* or *fidelity criterion*, and $E\{\cdot\}$ denotes the mathematical expectation. The distortion measure $d(x, y)$ quantifies the cost or the distortion resulting from reproducing x as y. The most commonly used fidelity criterion is the squared error, that is, $d(x, y) = (x - y)^2$.

The *quantization rate* R of a scalar quantizer is the number of bits required to represent the value x. It is measured in bits per sample.

The way of computing the quantization rate depends on the quantizer type. If each of the M possible cell numbers (for simplicity we assume that M is equal to a power of 2) is coded by a binary codeword of the same length, $\log_2 M$ bits, then such a quantizer is called a *fixed-rate* quantizer. The quantization rate in this case is equal to $R = \log_2 M$ bits/sample.

The goal of compression system design is to optimize the rate-distortion trade-off. Fixed-rate quantizers constrain this optimization by not allowing lossless coding of the quantizer outputs. It provides simpler encoding and allows buffering to be avoided which is required to match variable-length codewords to a fixed-rate digital channel.

Another scalar quantization technique is the so-called *variable-rate* quantization which uses lossless coding techniques to reduce the rate. If a quantizer is not constrained to be a fixed-rate, then its outputs can be lossless coded by a variable-length code. The numbers of cells at the output of the quantizer are considered as outputs of a discrete source with a given distribution. They are encoded by some variable-length code (for example, the Shannon code, the Huffman code, or arithmetic code).

Let $f(x)$ be a probability density function (pdf) of X. If $f(x)$ is known, then we can compute the probability of the ith reproduction level (the probability that x belongs to Δ_i) as

$$P(y_i) = \int_{t_{i-1}}^{t_i} f(x)\, dx. \tag{2.7}$$

The entropy of the quantizer outputs (the cell numbers) is equal to

$$H(Y) = -\sum_{i=1}^{M} P(y_i) \log_2 P(y_i).$$

The scalar quantizer with the outputs (the cell numbers) coded by a variable-length code according to the probability distribution $(P(y_1), P(y_2), \ldots, P(y_M))$ is called a *variable-rate* quantizer. In this case the number of cells M is not necessarily a power of 2 and R is the average rate or the expected number of bits required to specify the quantized value.

It is known from the theory of variable-length lossless coding that by using entropy coding methods such as, for example, arithmetic or enumerative coding, R can almost achieve $H(Y)$.

2.2.1   Uniform scalar quantization

A scalar quantizer is said to be *uniform* if the thresholds t_i are equally spaced or, in other words, if all cells (subintervals) have the same width, δ say, and the reproduction levels y_i are the midpoints of the cells.

If the outermost cells are finite, then the reproduction levels are:

$$y_i = \frac{t_{i-1} + t_i}{2} = t_{i-1} + \frac{\delta}{2}, \, i = 1, 2, \ldots, M.$$

Otherwise, if they are infinite, then for the outermost cells, $(t_0, t_1]$ and (t_{M-1}, t_M), the reproduction levels are equal to $y_1 = t_1 - \delta/2$ and $y_M = t_{M-1} + \delta/2$, respectively.

The value δ is called a *quantization step*. Assume for simplicity that we deal only with pdfs symmetric with respect to zero (any symmetric pdf can be reduced to this case by subtracting its mathematical expectation). Then one difference between fixed-rate and variable-rate uniform scalar quantizers is the way in which they exploit zero value. The fixed-rate quantizer uses zero value as a threshold, whereas for the variable-rate quantizer zero is a reproduction level. Also, the number of cells in these quantizers differ. The number of cells M for the fixed-rate quantizer is a power of 2. For the variable-rate quantizer M is always odd.

Example 2.1 A uniform fixed-rate quantizer for a source with bounded output ($x_{\min} = A = -1.0$ and $x_{\max} = B = 1.0$) is depicted in Fig. 2.4 as a collection of intervals bordered by thresholds ($t_0 = -1.0$, $t_1 = -0.5$, $t_2 = 0.0$, $t_3 = 0.5$, $t_4 = 1.0$) and with approximating values $y_1 = -0.75$, $y_2 = -0.25$, $y_3 = 0.25$, $y_4 = 0.75$. The rate of this quantizer is equal to 2 bits/sample.

The most commonly used uniform variable-rate quantizer is based on rounding off, that is, the cell number is computed as

$$i = \left[x/\delta \right]$$

where δ is the width of each cell and $[\cdot]$ denotes rounding. The approximating value corresponding to the input x is computed as

$$y = Q(x) = \delta \left[\frac{x}{\delta} \right]$$

that is, the ith cell is determined as $\Delta_i = (i\delta - \delta/2, i\delta + \delta/2]$, $y_i = i\delta$.

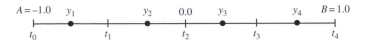

Figure 2.4   A uniform quantizer

Thus, for this quantizer the approximating values are

$$\ldots, -2\delta, -\delta, 0, \delta, 2\delta, \ldots$$

and the thresholds are

$$\ldots, -\frac{3\delta}{2}, -\frac{\delta}{2}, \frac{\delta}{2}, \frac{3\delta}{2}, \ldots.$$

It is evident that this quantizer is determined by only one parameter δ and does not require comparison of the input value x with thresholds.

Using binary codewords (binary representations of the numbers i) the dequantizer first recovers i and then recovers x as the approximating value y_i.

To estimate the quantization error it is necessary to choose a distortion measure. As mentioned above, the most commonly used criterion is the squared error, that is, $d(x, y) = (x - y)^2$. Consider the quantizer illustrated in Fig. 2.4. Let, for example, the value x be equal to 0.33. Then we obtain $i = 3$, $y_3 = 0.25$, the binary representation of the cell number is 10 (we count cells starting from 0). We spend 2 bits for storing x and the squared error is equal to 0.0064. If the pdf $f(x)$ of the random continuous variable X is unknown, then we estimate the average distortion measure as a sample average, when the quantizer is applied to a sequence of real data. If $f(x)$ is known, then the average distortion becomes an expectation

$$D(Q) = \int_{-\infty}^{\infty} f(x)d(x, Q(x)) \, dx = \sum_i \int_{t_{i-1}}^{t_i} f(x)d(x, y_i) \, dx.$$

If the distortion is measured by the squared error, then $D(Q)$ becomes the Mean Squared Error (MSE) and is computed by the formula

$$D(Q) = \int_{-\infty}^{\infty} f(x)(x - Q(x))^2 \, dx = \sum_i \int_{t_{i-1}}^{t_i} f(x)(x - y_i)^2 \, dx. \qquad (2.8)$$

2.2.2   Nonuniform optimal scalar quantization

It is intuitively clear that if we know the statistic parameters of the variable X we can use this information in order to improve the characteristics of the quantizer.

Depending on the type of quantizer there are two approaches to optimize the scalar quantizer.

Consider (2.8), which determines the MSE of the uniform scalar quantizer. Assume that we deal with a fixed-rate quantizer and M is a given finite number. Let for simplicity x belong to an interval of the real line, $[A, B]$ say. Then (2.8) can be rewritten as

$$D = \sum_{i=1}^{M} \int_{t_{i-1}}^{t_i} f(x)(x - y_i)^2 \, dx. \qquad (2.9)$$

For a given number of cells M, the MSE (2.9) depends on $2M - 1$ parameters. Namely, it depends on $M - 1$ thresholds $\{t_1, t_2, \ldots, t_{M-1}\}$ and M approximating values $\{y_i\}$. By optimizing these parameters it is possible to minimize the quantization error (2.9).

This problem is a standard problem of finding the minimum of a function of $2M - 1$ variables. We can rewrite (2.9) as

$$D = \int_{-\infty}^{\infty} f(x)x^2\, dx - 2\sum_{i=1}^{M} y_i \int_{t_{i-1}}^{t_i} f(x)x\, dx + \sum_{i=1}^{M} y_i^2 \int_{t_{i-1}}^{t_i} f(x)\, dx.$$

By differentiating (2.9) with respect to the y_i's and the t_i's and setting the derivatives equal to zero, we obtain

$$-2\int_{t_{i-1}}^{t_i} f(x)\, x\, dx + 2y_i \int_{t_{i-1}}^{t_i} f(x)\, dx = 0$$

$$-2y_i t_i f(t_i) + y_i^2 f(t_i) + 2y_{i+1} t_i f(t_i) - y_{i+1}^2 f(t_i) = 0.$$

Thus, under the condition that for all $x \in [A, B]$, $f(x) \neq 0$, the optimization problem has the following solution

$$t_i = (y_i + y_{i+1})/2 \tag{2.10}$$

$$y_i = \frac{m_i}{P(y_i)} \tag{2.11}$$

where

$$m_i = \int_{t_{i-1}}^{t_i} f(x)x\, dx \tag{2.12}$$

and where $P(y_i)$ determined by (2.7) is the probability that the random variable X gets the value x which belongs to the ith cell, and (2.11) determines the *centroid* of the area of the pdf $f(x)$ between t_{i-1} and t_i. Informally, centroid is the average of all points between t_{i-1} and t_i weighted by the corresponding values of $f(x)$. The quantization procedure described by formulas (2.10)–(2.11), is called *optimal nonuniform scalar quantization*. This quantization procedure was first described by S. P. Lloyd. Unfortunately, Lloyd's work was not published as a paper. Instead, it was presented at the 1957 Institute of Mathematical Statistics (IMS) meeting and appeared in print only as a Bell Laboratories Technical Memorandum. As a consequence, these results were not widely known in the engineering literature for many years, and they were independently rediscovered. In particular, this quantization technique became more popular in 1960 when it was rediscovered in (Max 1960). J. Max published another proof of the Lloyd optimality properties, and investigated numerically the design of fixed-rate quantizers for a variety of input pdfs. For example, he designed the fixed-rate quantizer for the case when the random variable X is Gaussian, that is, when its pdf $f(x)$ is

$$f(x) = \frac{1}{\sigma\sqrt{2\pi}} \exp\left(-(x-m)^2/(2\sigma^2)\right)$$

where m denotes the mathematical expectation and $\sigma^2 = \text{Var}(X)$ is the variance of X. Further, we always assume that $m = 0$.

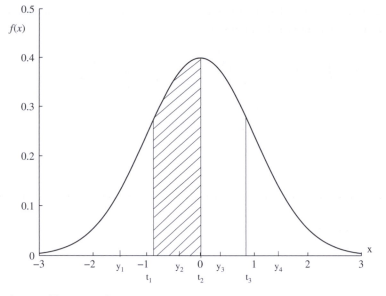

Figure 2.5   A nonuniform quantizer

Using (2.10)−(2.11) we obtain the following expression for the optimal MSE

$$D_{\text{opt}} = \int_{-\infty}^{\infty} x^2 f(x)\, dx$$

$$= E - \sum_{i=1}^{M} \frac{m_i^2}{P(y_i)} \tag{2.13}$$

where $E = \text{Var}(X)$ represents the average energy of random variable X with zero mean.

In Fig. 2.5 we show a pdf $f(x)$ with nonuniform quantizer (thresholds and approximating values) for the case $M = 4$. It follows from Fig. 2.5 that $t_i - y_i = y_{i+1} - t_i$ and the value y_2 is the centroid of the cell $(t_1, t_2]$ under the hatched area of $f(x)$.

The nonuniform optimal quantizer has cells of different width. The cell widths are small in the regions of highly probable values x and they are large in the regions of less probable values x.

In the general case, for nontrivial pdf $f(x)$, it is impossible to obtain explicit solutions of (2.10)−(2.11) analytically. In the next subsection we will consider an iterative procedure which determines the solution of (2.10)−(2.11) numerically.

If the nonuniform quantizer is not restricted to be a fixed-rate quantizer, then we can design the so-called *Entropy Constrained Scalar Quantizer* (ECSQ). Unlike nonuniform quantizers designed to minimize the distortion for a given number of cells, this quantizer has the optimal number of cells as well as the optimal set of approximating values and thresholds minimizing the average distortion with an entropy constraint, R_0 say. For this case the optimization problem reduces to the problem of finding the conditional minimum of the MSE, which can be formulated as follows:

$$D(Q, R) = \min_{\{t_i\},\{y_i\},M} \left\{ \sum_{i=1}^{M} \int_{t_{i-1}}^{t_i} (x - y_i)^2 f(x)\, dx \right\} \quad H(\{t_i\}, M) = R_0$$

where $\{t_i\}$ and $\{y_i\}$ denote the sets of thresholds and approximating values, respectively.

This problem can be solved by Lagrangian minimization. To do that we construct the Lagrangian function

$$L = \left\{ \sum_{i=1}^{M} \int_{t_{i-1}}^{t_i} (x - y_i)^2 f(x)\, dx \right\} + \lambda \varphi \left(\{t_i\}, M \right)$$

where

$$\varphi \left(\{t_i\}, M \right) = H \left(\{t_i\}, M \right) - R_0$$

and search for the minimum of L over $\{t_i\}$, $\{y_i\}$, M. The problem of constructing the optimal entropy constrained quantizer is solved for memoryless sources (see, for example, Noll and Zelinski [1978]; Farvardin and Modestino [1984]). The main shortcoming of the ECSQ is the high computational complexity of the design procedure. Notice also that to describe any optimal nonuniform quantizer, a fixed-rate or variable-rate one, we need to specify the set of thresholds $\{t_i\}$ and the set of approximating values $\{y_i\}$.

It is evident that the most attractive quantizer from the implementation point of view is the uniform scalar quantizer based on rounding off. On the other hand, the best $R(D)$ function provides the ECSQ. There are a few approaches to improve performances of the uniform quantizer while keeping its low computational complexity. The basic idea of all approaches is to optimize the thresholds and approximating values independently.

One approach implies that we choose the thresholds uniformly, that is,

$$t_i = \delta \left(i + \frac{1}{2} \right)$$

and let the approximating values satisfy (2.11). The corresponding quantizer is called the *Optimal Uniform Quantizer* (OUQ).

Notice that in some papers OUQ implies the uniform quantizer with all cells of the same width except for the two outermost cells which are semi-infinite. Let the number of cells be equal to M and all cells except the two outermost have the same width δ, and let the approximating values be $y_i = t_i - \delta/2$. Then the MSE (2.9) depends only on two parameters: the cell width δ and one of the thresholds, t_1, say. By optimizing these two parameters in order to minimize (2.9), we obtain the *optimal uniform quantizer*.

Another approach leads to the *Quantizer with Extended Zero Zone* (QEZZ). This quantizer has a set of thresholds computed by the formula

$$t(j, \alpha) = \{ \pm \alpha 2^{j-1}, \pm \alpha \left(2^{j-1} + 1 \right), \pm \alpha \left(2^{j-1} + 2 \right), \dots \},\ \alpha > 0\,,\ j = 0, 1, \dots$$

The approximating values are chosen as midpoints of the cells. It is easy to see that $t(0, \alpha)$ describes the set of thresholds $\pm \alpha/2, \pm 3\alpha/2, \dots$ of the uniform scalar quantizer.

If the approximating values of the quantizer with an extended zero zone are computed by (2.11), then the quantizer is called *Optimal Quantizer with Extended Zero Zone*

(OQEZ). Often, in practice, we use the quantizer with extended zero zone with only the two closest to zero approximating values chosen according to (2.11). The other approximating values are midpoints of the cells. Such a quantizer is called the *Suboptimal Quantizer with Extended Zero Zone* (SQEZ). In Section 3, we discuss the performances of the considered scalar quantizers.

2.2.3   The Lloyd–Max procedure

Even if the pdf $f(x)$ is known, the method of solving (2.10) and (2.11) represents an iterative procedure. Usually we start with the uniform scalar quantization and then compute the new approximating value using (2.11). If after that the quantization error is reduced, we compute new thresholds. We continue these computations until the reduction of the quantization error becomes negligible. This procedure was also suggested by Lloyd but due to the above-mentioned reasons it is usually called the Lloyd–Max quantization procedure.

Let the required quantization rate be equal to R_0 and let the input pdf $f(x)$ be known. Choose the accuracy $\epsilon > 0$ and/or the maximal number of iterations N_{I}. The Lloyd–Max procedure is presented as Algorithm 2.1.

Input:
quantization rate R_0
pdf $f(x)$
accuracy $\epsilon > 0$
maximal number of iterations N_{I}
average sample energy $E = \mathrm{Var}(X)$
Output: approximating values y_i, $i = 1, 2, \ldots, M$
thresholds t_i, $i = 1, 2, \ldots, M - 1$;

Initialization:
Compute the number of cells $M = 2^{R_0}$
Set the current error value $D_c = E$
Set $D = 0$
Set the current number of iterations $N = 0$
Compute thresholds t_i, $i = 1, 2, \ldots, M - 1$ for the uniform quantizer

while $N < N_{\mathrm{I}}$ **and** $D_c - D > \epsilon$ **do**

1. Compute the approximating values y_i, $i = 1, 2, \ldots, M$ according to (2.11) and the new error value according to (2.13).
2. Compute the thresholds t_i, $i = 1, 2, \ldots, M - 1$, according to (2.10).
3. $D_c \leftarrow D$; $N \leftarrow N + 1$.

end

Algorithm 2.1   Lloyd–Max procedure

If $f(x)$ is unknown, then the same iterative procedure can be applied to a sequence $x_1, x_2, \ldots, x_k$ of values of the random variable X observed at the quantizer input during k sequential time moments. In this case the approximating values are determined by the formula

$$y_i = \frac{\hat{m}_i}{\hat{P}(y_i)}$$

where $\hat{m}_i = k^{-1} \sum_{x \in [t_{i-1}, t_i)} x$, $\hat{P}(y_i) = k_i/k$ and k_i is the number of values x which lie in the ith cell. Thus, we have

$$y_i = \frac{\sum_{x \in (t_{i-1}, t_i]} x}{k_i}$$

as a sample average of values x_i, which lie in the ith cell. It is evident that $\hat{m}_i$ and $\hat{P}(y_i)$ are estimates of m_i and $P(y_i)$ is obtained for the observed sequence $x_1, x_2, \ldots, x_k$. It is easy to see that the Lloyd–Max procedure results in reducing those intervals, which contain many values x_i and extending those intervals, which contain a small number of values x_i.

Thus we obtain an optimal partition of the real line into intervals (cells) and a set of approximating values that are the centroids of these cells. Using the obtained thresholds t_i and approximating values y_i we can perform the nonuniform scalar quantization of the random variable X. Notice that in the general case (if the distortion measure is a convex function but not necessarily the squared error) we start with a set of approximating values, find the optimal set of cells (the nearest neighbor regions with respect to the distortion measure), then find the optimal approximating values for these regions, and repeat the iteration for this new set of approximating values. The expected distortion is decreased at each stage in the algorithm, so it will converge to a local minimum of the distortion.

2.3 Vector quantization

Vector quantization is a generalization of scalar quantization to quantization of a vector or point in the n-dimensional Euclidean space $\mathcal{R}^n$ or in its subspace. Unlike scalar quantization, which is mainly used for analog to digital conversion, vector quantization is often used in the cases when the input signal is already digital and the output is a compressed version of the original signal.

Let a random variable take values x from the set $X \subseteq \mathcal{R}$. Let $X^n \subseteq \mathcal{R}^n$ denote a set of vectors $\boldsymbol{x} = (x_1, x_2, \ldots, x_n)$, where x_i is a variable value at time i or, in other words, $\boldsymbol{x}$ represents a vector of variable values for n consecutive time moments.

- Vector quantization is a mapping Q of an input vector $\boldsymbol{x} = (x_1, x_2, \ldots, x_n)$ from the set X^n into the closest (with respect to the chosen distortion measure) approximating vector $\boldsymbol{y} = (y_1, y_2, \ldots, y_n)$ from the discrete *approximating set* $Y = \{\boldsymbol{y}_1, \boldsymbol{y}_2, \ldots,$

y_M}, the parameter n is called *dimension* of the quantizer. The approximating set Y is often called *codebook*.

- To construct a vector quantizer means to split X^n into M areas or cells S_i, $i = 1, 2, \ldots, M$, such that $\bigcup_i S_i = X^n$, $S_i \cap S_j = \emptyset$, $i \neq j$, and $y_i \in S_i$.

The quality of the quantizer is usually measured by the average quantization error due to replacing x by $y = Q(x)$ and it is defined as

$$D_n(Q) = \mathrm{E}\{d\,(x, Q(x))\}. \tag{2.14}$$

If the distortion measure $d(x, Q(x))$ is the normalized squared error or normalized squared Euclidean distance between the input vector x and the approximating vector y, that is,

$$d(x, y) = \frac{1}{n}d_{\mathrm{E}}^2(x, y) = \frac{1}{n}(x - y)(x - y)^T = \frac{1}{n}\sum_{i=1}^{n}(x_i - y_i)^2 = \frac{1}{n}\|x - y\|^2$$

then (2.14) can be rewritten as

$$D_n(Q) = \frac{1}{n}\mathrm{E}\{\|x - Q(x)\|^2\} = \frac{1}{n}\sum_{i=1}^{M}\mathrm{E}\{\|x - y_i\|^2\}. \tag{2.15}$$

Assume that the n-dimensional pdf $f(x)$ for the set X^n is known, then (2.15) has the form

$$D_n(Q) = \frac{1}{n}\sum_{i=1}^{M}\int_{S_i} f(x)\|x - y_i\|^2\,dx \tag{2.16}$$

and the probability of the vector y_i is

$$P(y_i) = \int_{S_i} f(x)\,dx.$$

The *quantization rate* is the number of bits required to represent the vector x per quantizer dimension n. For the fixed-rate quantizer it is determined as follows

$$R = \frac{\log_2 M}{n} \text{ bits/sample.} \tag{2.17}$$

For the variable-rate vector quantizer, by using the standard lossless coding technique we can achieve

$$R = -\frac{1}{n}\sum_{j=1}^{M} P(y_j)\log_2 P(y_j). \tag{2.18}$$

Analogously to the scalar quantizer any vector quantizer is characterized by its *rate-distortion function* $R(D_n)$.

Example 2.2 Let the approximating set Y contain 4 vectors of length 2:

$$y_1 = (0.2, 0.3)$$

$$y_2 = (0.3, 0.1)$$

$$y_3 = (0.1, 0.5)$$

$$y_4 = (0.0, 0.4).$$

Assume that we quantize the vector $x = (0.18, 0.25)$ and that the distortion measure is the squared error; then the closest approximating vector is $y_1 = (0.2, 0.3)$. The quantization error is equal to

$$d(x, y) = \frac{1}{2}((0.2 - 0.18)^2 + (0.3 - 0.25)^2) = 1.45 \cdot 10^{-3}$$

and the quantization rate is equal to $R = (\log_2 4)/2 = 1$ bit/sample.

2.3.1 Optimal vector quantization with a given size of codebook. The Linde–Buzo–Gray procedure

In this subsection we consider a vector quantizer which is a generalization of the optimal nonuniform fixed-rate scalar quantizer. Assume that $f(x)$ is known and we would like to minimize the quantization error (2.16) under the condition that the size of the codebook is fixed. Then, differentiating the right-hand side of (2.16) with respect to y_i and setting derivatives to zero, we obtain

$$y_i = \frac{\int_{S_i} x f(x) \, dx}{\int_{S_i} f(x) \, dx}, i = 1, 2, \ldots, M. \tag{2.19}$$

It is easy to see that (2.19) describes the centroids of the regions S_i, $i = 1, 2, \ldots, M$.

To specify the cells S_i, $i = 1, 2, \ldots, M$, we introduce the following definitions:

- Consider a discrete set $P = \{p_1, p_2, \ldots, \}$, where $p_i \in \mathcal{R}^n$. It is known that each point $p_i \in \mathcal{R}^n$ is a centroid of its *Voronoi region* or, in other words, an area of $\mathcal{R}^n$ which contains those points of $\mathcal{R}^n$ which are at least as close to p_i as to each other point p_j:

$$V_i = \{x \in \mathcal{R}^n : d(x, p_i) \le d(x, p_j), \text{ for all } j \ne i\}.$$

- The Voronoi regions never intersect (except at the boundaries). As noticed in Gersho and Gray (1992) to avoid ambiguity we can always assign x to be a member of V_m where m is the smallest index i for which $d(x, p_i)$ attains its minimum value. The union of the Voronoi regions coincides with $\mathcal{R}^n$.

- We say that V_i is *convex* if from $x_1 \in V_i$ and $x_2 \in V_i$ it follows that $ax_1 + (1 - a)x_2 \in V_i$ for all $0 < a < 1$.

Thus, the so-called *Voronoi partition* generates such cells V_i that each of them consists of all points x which have less distortion when reproduced with p_i than with any other vector from P. Notice that in the general case the Voronoi cells are not polytopal nor even convex but if the distortion measure is the squared error, then the Voronoi cells are polytopal. Moreover, in this case the borders of the cells can be specified from the values of vectors p_i.

Now consider the rule of finding the approximating vector y_i for an input vector x. The squared quantization error for a given input vector x will be minimal if we choose as an approximation such a vector $y_i \in Y$ that minimizes $\|x - y_i\|^2$, that is, y_i satisfies the inequality

$$\|x - y_i\|^2 \le \|x - y_j\|^2, \text{ for all } j \ne i. \tag{2.20}$$

It is evident that the quantizer with such a decision rule performs the Voronoi partition. In this case the border between S_i and S_j is a hyperplane perpendicular to the segment connecting y_i and y_j. It can be described as

$$H_{ij} = \left\{ x : (x, y_j - y_i) + \frac{1}{2}(\|y_i\|^2 - \|y_j\|^2) = 0 \right\}.$$

Every hyperplane that determines a border of a cell is characterized by two approximating vectors to which this hyperplane is equidistant. Thus, the quantization error is, in fact, determined by choosing the set of approximating vectors $\{y_i\}$ since this set uniquely determines the shapes of the quantization cells. Notice that if pdf $f(x)$ is unknown, then in order to find the centroids of the Voronoi polyhedrons we can use a training sequence $x_1 = (x_{11}, x_{12}, \ldots, x_{1n})$, $x_2 = (x_{21}, x_{22}, \ldots, x_{2n})$, $\ldots$, $x_k = (x_{k1}, x_{k2}, \ldots, x_{kn})$, observed at the input of the quantizer. For the squared error measure the centroid reduces to the arithmetic average

$$\hat{y}_i = \frac{1}{k_i} \sum_{x \in S_i} x$$

where k_i is the cardinality of the set S_i. Although it is impossible to find a closed form solution of the optimization problem, formulas (2.19) and (2.20) can be used to organize an iterative procedure which allows us to find the solution numerically. This procedure, which is often called the Linde–Buzo–Gray (LBG) algorithm (Linde *et al.* 1980), represents a generalization of the Lloyd–Max procedure to the n-dimensional case. Starting with an arbitrary initial codebook in each iteration, we construct the quantization cells according to (2.20) and then modify the codebook according to (2.19). We stop when the reduction of the quantization error becomes negligible.

Next, we consider the LBG procedure in more detail. Let the required quantization rate be equal to R_0. We assume that the size of the codebook $M = 2^{R_0 n}$ is significantly less than the number of observed vectors k. Choose the accuracy $\epsilon > 0$ and the maximal number of iterations N_I. The LBG procedure is presented as Algorithm 2.2.

Input:

quantization rate R_0

accuracy $\epsilon > 0$

maximal number of iterations N_I

number of observed vectors k

set of the observed vectors $\{x_j\}$, $j = 1, \ldots, k$ of length n

Output: approximating vectors y_i, $i = 1, 2, \ldots, M$

Initialization:

compute $M = 2^{R_0}$

Initialize the codebook by choosing as approximating vectors y_i, $i = 1, 2, \ldots, M$, arbitrary vectors from $\{x_j\}$, $j = 1, \ldots, k$.

Set the previous step error $D_p = \infty$

Set the current error $D_c = 0$

Set the current number of iterations $N = 0$

while $N < N_I$ **and** $D_p - D_c > \epsilon$ **do**

1. Set the current error $D_c = 0$,
   $s_i = (0, \ldots, 0)$ (the all-zero vector of length n),
   counters $k_i = 0$, $i = 1, 2, \ldots, M$.
   foreach x_j, $j = 1, 2, \ldots, k$ **do**
   - Find the closest approximating vector y_i according to (2.20).
   - Modify the componentwise sum of vectors x_j closest to y_i,
     $s_i \leftarrow s_i + x_j$.
   - Increment the number of vectors x_j closest to y_i, that is,

     $$k_i \leftarrow k_i + 1.$$

   - Update the value of the quantization error

     $$D_c \leftarrow D_c + \frac{1}{n}\|x_j - y_i\|^2.$$

   end
2. Modify the codebook according to

   $$y_i \leftarrow s_i/k_i, \quad i = 1, 2, \ldots, M.$$

3. Update the average error $D_c \leftarrow D_c/k$.
4. Increment the number of iterations $N \leftarrow N + 1$.

end

Algorithm 2.2   LBG algorithm

The LBG procedure results in constructing a codebook of approximating vectors for a random input vector. Analogously to the Lloyd–Max procedure, this procedure displaces n-dimensional approximating vectors into highly probable regions according to the n-dimensional pdf of the input vector. Vector quantization unlike scalar quantization has cells which are not intervals of the real line and have not even the same shape.

When the codebook is found, the quantization procedure itself consists in finding for any input vector x one of the vectors y_i, $i = 1, 2, \ldots, M$ closest to x in terms of the squared error $\|x - y\|^2$.

The main shortcoming of vector quantization is its high computational complexity. To find the closest approximating vector it is necessary to perform an exhaustive search among $M = 2^{nR_0}$ vectors. For example, if $R_0 = 0.5$ bit/sample and $n = 32$, then the size of the codebook is equal to $M = 2^{16}$. It is possible to reduce the computational complexity of vector quantization by using structured codebooks.

2.3.2   Lattice vector quantizers

As mentioned before, the LBG procedure yields unstructured codebooks and the corresponding quantization procedure in this case typically represents an exhaustive search over the set of approximating vectors. The codebooks of a special class of vector quantizer called *lattice vector quantizers* have a highly regular structure. By using this regularity it is possible to simplify the quantization procedure. In this section we start with some basic facts about lattices. The thorough overview of lattice theory and applications can be found in Conway and Sloane (1988).

Let us introduce some definitions.

- Let $\{u_1, \ldots, u_n\}$ be a set of linearly independent vectors in $\mathcal{R}^n$. Then the *lattice* Λ generated by $\{u_i\}$, $i = 1, 2, \ldots, n$, is the set of all points of the form $y = \sum_{i=1}^{n} c_i u_i$, where c_i are all integers. The vectors $\{u_i\}$ represent a *basis* for the n-dimensional lattice.
- The matrix

$$U = \begin{pmatrix} u_1 \\ \ldots \\ u_n \end{pmatrix}$$

  is called the *generator matrix* of the lattice, that is, any vector of the lattice can be written as $y = cU$, $c = (c_1, c_2, \ldots, c_n)$, c_i are integers.

Example 2.3   The simplest n-dimensional lattice is the integer or *cubic* lattice denoted by Z^n, which consists of all vectors of dimension n with integer coordinates. Its generator matrix is

$$U = I$$

where I denotes the identity matrix of size $n \times n$. If $n = 1$ the corresponding one-dimensional (1D) lattice Z is the set of integers.

If $n = 2$, we obtain the *square lattice* Z^2. Its generator matrix is

$$U = \begin{pmatrix} 1 & 0 \\ 0 & 1 \end{pmatrix}.$$

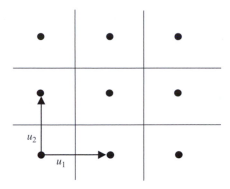

Figure 2.6 Voronoi cells and basis vectors of the square lattice

Any vector $\boldsymbol{y} = (y_1, y_2)$ of this lattice can be computed as

$$y_1 = c_1$$

$$y_2 = c_2$$

where c_1, c_2 are integers.

The fragment of the square lattice, its basis vectors, and the Voronoi cells are shown in Fig. 2.6.

Example 2.4 The so-called *hexagonal* lattice A_2 can be given by its generator matrix:

$$U = \begin{pmatrix} 1 & 0 \\ 1/2 & \sqrt{3}/2 \end{pmatrix}.$$

Any vector $\boldsymbol{y} = (y_1, y_2)$ of this lattice can be computed as

$$y_1 = c_1 + c_2/2$$

$$y_2 = c_2\sqrt{3}/2$$

where c_1, c_2 are integers.

Figure 2.7 shows an example of basis vectors and Voronoi cells of the hexagonal lattice.

We can say that a lattice Λ is an infinite regular array which covers $\mathcal{R}^n$ uniformly. Lattice quantization uses the points of a lattice or some subset of a lattice as approximating vectors for vector quantization. In particular, the points of the lattice Z can be used as approximating values of the uniform scalar quantizer. In other words, a *lattice quantizer* is a vector quantizer whose codebook is either a lattice (if the codebook is infinite) or a truncated lattice (if the size of the codebook is finite).

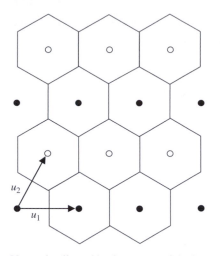

Figure 2.7   Voronoi cells and basis vectors of the hexagonal lattice

The lattice points (except the outermost points of the truncated lattice) are centroids of the Voronoi polyhedrons which are all of the same size and shape, that is, they are *congruent*. For simplicity (to avoid a separate study of Voronoi regions of the outermost points) we consider infinite codebooks in the sequel, that is, all Voronoi regions are supposed to be congruent without exceptions. One more important property of any lattice is that it always contains the all-zero point or the origin of coordinates.

Since all Voronoi regions are congruent we usually consider only the Voronoi region which contains the origin of coordinates $V_0(\Lambda)$. For example, the corresponding Voronoi region of Z^n contains all vectors of dimension n which lie within a cubic region of unit volume centered at the origin. The translates $V_0(\Lambda) + y$ of $V_0(\Lambda)$ for all $y \in \Lambda$ cover $\mathcal{R}^n$. Infinite codebook implies that the outputs of the quantizer are entropy encoded, we deal with variable-rate quantizer and a finite codebook usually corresponds to a fixed-rate quantizer. The quantization rate is determined by (2.18) and (2.17), respectively.

Surely, the algorithms for finding the closest point (approximating vector) in the lattice depend on the lattice structure. The simplest lattice quantizer can be constructed by using n uniform scalar quantizers. In other words, if the codebook is the n-dimensional integer lattice Z^n, then the closest approximating vector $y = (y_1, y_2, \ldots, y_n)$ for an input vector $x = (x_1, x_2, \ldots, x_n)$ can be found by rounding off each of n components x_i. Since vector quantization is performed by separately quantizing the components of x, the average quantization error coincides with the average quantization error for the corresponding scalar quantizer. The Voronoi cells of such a vector quantizer are n-dimensional cubes. In the general case lattice quantizers have nonrectangular Voronoi regions. In the next chapter we discuss the so-called *shaping gain* that is the gain in the average distortion obtained for lattice quantizers compared to scalar quantizers by manipulating the shape of the Voronoi cells. Notice that spherical Voronoi regions maximize the shaping gain but they cannot cover $\mathcal{R}^n$.

To demonstrate the advantage of lattice quantizers in terms of computational complexity we consider the following example.

Example 2.5 Assume that we quantize a real vector

$$x = (x_1, x_2) = (-5.6, 0.82)$$

using the hexagonal lattice A_2. Denote by S^n a *scaled integer lattice* (all vectors of dimension n with integer components, where each component is scaled by a real coefficient). Sometimes, instead of S^n we will use the notation aZ^n. The term a in front of Z^n means that the lattice Z^n is scaled by factor of a in all directions. It is easy to see that the lattice A_2 can be represented as a union of the scaled square lattice S^2 (filled points in Fig. 2.7) and its translate by vector $(1/2, \sqrt{3}/2)$ (unfilled points in Fig. 2.7). The points of S^2 have the form:

$$y_1 = \left(c_1, c_2\sqrt{3} \right) \tag{2.21}$$

where c_1, c_2 are integers and the points of its translate are

$$y_2 = \left(c_1 + \frac{1}{2}, c_2\sqrt{3} + \frac{\sqrt{3}}{2} \right) = y_1 + \left(\frac{1}{2}, \frac{\sqrt{3}}{2} \right). \tag{2.22}$$

First we find for x the closest point in S^2. As explained above, this is equivalent to performing scalar quantization for each component (the quantization step for each component is equal to the corresponding scaling coefficient). For the two-dimensional (2D) input vector $(-5.6, 0.82)$ in the scaled lattice S^2, that is among points in the form (2.21), we obtain the approximation $(-6, 0)$. Then we find the closest point in the translate of S^2 (2.22) and obtain the following approximation: $(-5.5, \sqrt{3}/2)$. Comparing the quantization errors $\|x - y\|^2/2$ for these two approximations, we found that the best approximation is $(-5.5, \sqrt{3}/2)$ and the corresponding quantization error is equal to 0.0058. The output of the quantizer is $c_1 = -6$, $c_2 = 1$.

In the considered example, instead of searching among all possible approximating 2D vectors, we reduced the quantization procedure to four scalar quantizations in 1D lattices and chose the best approximation.

There exist several algorithms for finding the closest point in the specific lattice, taking the lattice structure into account (see, for example, Conway and Sloane [1988]). A special class of lattices are those based on linear codes. In order to simplify the quantization procedure for lattices of this type we can apply decoding techniques well developed in the area of error-correcting codes (see, for example, Lin and Costello [2004] and references herein). Let $C = \{c_i\}$ be an (n, k) binary linear code, then the lattice $\Lambda(C)$ is defined by

$$\Lambda(C) = \{y \in Z^n | y \equiv c(\text{mod } 2) \text{ for some } c \in C\}.$$

Assume that the generator matrix G of the rate $R_c = k/n$ code $\mathcal{C}$ is systematic and given as

$$G = (I|B)$$

where I is an identity matrix of size $k \times k$ and B is a matrix of size $k \times (n - k)$. Then the generator matrix of the lattice $\Lambda(\mathcal{C})$ has the form

$$U = \begin{pmatrix} I & B \\ 0 & 2I \end{pmatrix}.$$

Therefore, the lattice $\Lambda(\mathcal{C})$ contains all points from $2Z^n$ (all n-dimensional vectors with even integer components) and those points from Z^n which are congruent to some codeword modulo 2. We can consider the lattice $\Lambda(\mathcal{C})$ as a union of 2^k translates (*cosets*) of the sublattice $2Z^n$:

$$\Lambda(\mathcal{C}) = \bigcup_{i=0}^{2^k-1} \left(c_i + 2Z^n \right).$$

The described structure of the lattice can be used in order to reduce the vector quantization procedure to a few scalar quantizations of the input vector components. The main idea is to scalar quantize the properly scaled input vector in each of 2^k possible cosets of $2Z^n$. The scaling coefficient plays the role of the quantization step of the scalar quantizer. The output of the quantizer is the best, in terms of the average quantization error, quantized sequence, and the number of the corresponding coset. Since each coset is determined by a codeword, then the chosen coset number is simply the number of the corresponding codeword.

The computation complexity of such vector quantization does not depend on the quantization rate. It depends only on the length of the input vector and the number of possible cosets of $2Z^n$ that are equal to 2^k.

Now we are ready to formulate the quantization procedure for the (scaled) input vector x:

Input:
input vector x
Output:
pair (q_i, c_i) minimizing the quantization error

foreach *codeword* c_i, $i = 0, 1, \ldots, 2^k - 1$ **do**

- Subtract the corresponding codeword c_i from the input vector x and obtain $d_i = x - c_i$.
- Quantize each component of d_i by the uniform scalar quantizer with quantization step 2 and obtain the quantized vector q_i.
- Compute the quantization error $\|x - y_i\|^2$, where $y_i = 2q_i + c_i$.

end
Find the best pair (q_i, c_i) minimizing the quantization error.

Algorithm 2.3   Lattice quantization

Thus, instead of x we store or transmit q_i and i.

A dequantizer performs the following steps:

Input:
pair (q_i, c_i)
Output:
approximating vector y

- Reconstruct the approximating vector $\hat{d}_i$ from q_i.
- Add to the approximating vector the corresponding codeword:
$$y = \hat{d}_i + c_i.$$

Algorithm 2.4 Lattice dequantization

Example 2.6 Let C be a $(3, 2)$ linear block code with the generator matrix

$$G = \begin{pmatrix} 1 & 0 & 1 \\ 0 & 1 & 1 \end{pmatrix}.$$

Then the generator matrix of the lattice based on C is

$$U = \begin{pmatrix} 1 & 0 & 1 \\ 0 & 1 & 1 \\ 0 & 0 & 2 \end{pmatrix}.$$

Assume that the input vector is $x = (0.4, 1.2, 3.7)$. For each of the codewords

$$c_0 = (0, 0, 0)$$

$$c_1 = (0, 1, 1)$$

$$c_2 = (1, 0, 1)$$

$$c_3 = (1, 1, 0)$$

we obtain the following quantized vectors:

$$q_0 = (0, 1, 2)$$

$$q_1 = (0, 0, 1)$$

$$q_2 = (0, 1, 1)$$

$$q_3 = (0, 0, 2).$$

The corresponding approximating vectors are:

$$\hat{\pmb{d}}_0 = (0, 2, 4)$$

$$\hat{\pmb{d}}_1 = (0, 0, 2)$$

$$\hat{\pmb{d}}_2 = (0, 2, 2)$$

$$\hat{\pmb{d}}_3 = (0, 0, 4).$$

By adding codewords to the approximating vectors, finally we obtain

$$\pmb{y}_0 = (0, 2, 4)$$

$$\pmb{y}_1 = (0, 1, 3)$$

$$\pmb{y}_2 = (1, 2, 3)$$

$$\pmb{y}_3 = (1, 1, 4).$$

The corresponding quantization errors are: 0.297, 0.230, 0.497, and 0.163. Hence, the closest coset is generated by the codeword $(1, 1, 0)$ and the best approximation is $(1, 1, 4)$. The output of the quantizer is the quantized vector $(0, 0, 2)$ and the codeword $(1, 1, 0)$, that is, $i = 3$.

To reduce further the computational complexity of the considered lattice quantization procedure, we will show that it can be interpreted as a procedure for choosing the best sequence of n scalar approximating values among the 2^k allowed sequences, where each approximating value is generated by one of the two scalar quantizers. The first scalar quantizer has the approximating values $\ldots, -4, -2, 0, 2, 4, \ldots$ and the second scalar quantizer has the approximating values $\ldots, -3, -1, 1, 3, \ldots$ We can say that the quantizer finds the best (in terms of the MSE) quantized sequence by switching between $2Z$ and $Z \setminus 2Z$, i.e. sets of even and odd integers.

Our quantization procedure consists in searching for the best approximation $\pmb{y}$ closest to the input vector $\pmb{x}$ in terms of MSE:

$$\min_{\pmb{y} \in \Lambda(C)} d(\pmb{x}, \pmb{y}) = \frac{1}{n}||\pmb{x} - \pmb{y}||^2 = \frac{1}{n} \sum_{t=1}^{n} (x_t - y_t)^2 \qquad (2.23)$$

$$= \frac{1}{n} \sum_{t=1}^{n} d(x_t, y_t).$$

Each approximating vector y belongs either to $2Z^n$ or to its translate $2Z^n + c$, where $c \in C$ is a codeword. It is determined as

$$y = 2i + c$$

where

$$i = \left[\frac{x - c}{2}\right]$$

is the sequence of n cell numbers describing the point closest to $x - c$ in $2Z^n$ and $[\cdot]$ denotes componentwise rounding off.

Taking into account that y is uniquely determined by c and i, we can rewrite (2.23) in the form

$$\min_{y \in \Lambda(C)} d(x, y(c)) = \min_{c \in C} d(c) = \sum_{t=1}^{n} d(c_t)$$

where

$$d(c_t) = (x_t - y_t(c_t))^2$$

$$y_t(c_t) = 2i_t + c_t \tag{2.24}$$

and

$$i_t(c_t) = \left[\frac{(x_t - c_t)}{2}\right]. \tag{2.25}$$

It is easy to see from (2.24) and (2.25) that each component $y_t(c_t)$ can be found by uniform scalar quantization in $2Z$ if $c_t = 0$ or in $Z \setminus 2Z$ if $c_t = 1$.

We come to the following quantization procedure:

Input:
input sequence $x = (x_1, x_2, \ldots, x_n)$
Output:
number m of the chosen codeword $c_m = (c_1^m, c_2^m, \ldots, c_n^m)$ and the sequence
$i_t(c_t^m)$, $t = 1, 2, \ldots, n$, $m = 0, 1, \ldots, 2^k - 1$

for $t = 1$ to n and $c = 0, 1$ **do**

   • $i_t(c) = round((x_t - c)/2)$
   • $d_t(c) = (x_t - 2i_t(c) - c)^2$

end
Find $c_m \in C$, $m = 0, 1, \ldots, 2^k - 1$ such that $d(c_m) = \sum_{t=1}^{n} d_t(c_t^m)$ is minimal.

Algorithm 2.5   Modified lattice quantization

The presented quantization procedure consists of two parts: computing cell numbers $i_t(c)$ and distortion measures $d_t(c)$, and search for the codeword best in the sense of

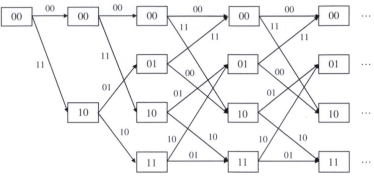

Figure 2.8   Code trellis

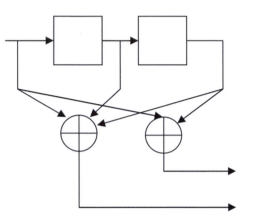

Figure 2.9   Encoder of convolutional code

the given distortion measure. In other words, finding the quantized vector implies an exhaustive search among the 2^k codewords. The computational complexity of this procedure can be reduced by using a trellis representation of the codes. In this case the search procedure can be implemented as the *Viterbi decoding algorithm* (Viterbi 1967). Binary convolutional codes (see, for example, Viterbi and Omura [1979]) look like a reasonable choice, since they determine infinite binary lattices.

Although we assume that the reader has encountered convolutional codes, we briefly discuss the notion "*code trellis*" and recall the Viterbi decoding algorithm.

In Fig. 2.8 a few levels of a trellis diagram (code trellis) are shown. The code trellis is generated by the encoder of an $R_c = 1/2$ convolutional code with constraint length $v = 2$ as shown in Fig. 2.9.

If the encoder is regarded as a finite state machine, then the nodes of the trellis correspond to the states of this machine. The constraint length v of the convolutional code is equal to the total number of delay elements of its encoder; that is, the number of states at each level of the trellis diagram is equal to 2^v. In general, the encoder of a rate $R_c = K/N$ convolutional code consists of K registers connected with N modulo 2 adders. We distinguish between constraint length v of the code equal to the total number of delay elements in the encoder registers and code

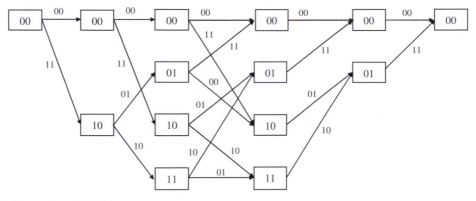

Figure 2.10 The terminated trellis

memory m equal to the number of delay elements in the longest of encoder registers. For rate $R_c = 1/N$ convolutional codes these two values coincide, i.e. $v = m$. Transitions between the states are shown by branches labeled by automata outputs generated by these transitions. For a convolutional code of rate K/N, each of the 2^K possible transitions determines N code symbols (trellis branch) corresponding to the given time moment. For rate $R_c = 1/2$, one input bit enters the encoder at each time moment and two output bits, depending on the current encoder state and the input bit, leave the encoder. Each path through the trellis corresponds to a codeword of infinite length.

Although the input and output sequences of a convolutional encoder are semi-infinite, in practice we deal only with terminated convolutional codes. An input sequence of length, Kh say, is followed by v zeros in order to force the encoder into zero state. The generated codeword has the length $n = N(h + v)$. The *zero-tail terminated* convolutional code obtained in such a way has the rate

$$R_{ct} = \frac{Kh}{N(h+v)} < R_c = \frac{K}{N}.$$

However, if h is large enough the loss in code rate is negligible. In Fig. 2.10 the terminated trellis is shown.

Assume that our lattice quantizer is based on such a terminated convolutional code. We already showed that the most computationally expensive part of the quantization procedure is the exhaustive search for the best codeword. However, for a code represented by its trellis it can be performed with complexity proportional to the number of states at each level of the trellis. The Viterbi algorithm for finding the best in the sense of a given additive metric[1] codeword is presented below as Algorithm 2.6. Notice that in fact this algorithm is the algorithm of finding the shortest path in the directed graph.

---

[1] For the quantization procedure the metric coincides with a distortion measure, in particular, with the squared quantization error.

Input:
input vector $x = (x_1, x_2, \ldots, x_n)$
Output:
information bits corresponding to survived path to the all-zero state
at level $h + v$.

Initialization: at time unit $t = 0$ set metric of the all-zero state to zero and assign
empty path to this node
for $t = 1$ **to** $t + h$ **do**

    foreach *state of 2^v states* **do**

- Compute the metric for two paths entering state by adding
  the entering branch metric to the metric of the previous state.
- Compare the metrics of two entering paths, select the path with
  the best (smallest) metric, and assign its metric to the given
  state.
- Compute the survived path to each state by appending the
  information symbol determining the transition from the
  previous state to the given state to the path entering
  the previous state.

    end

end
Output the survived path to the all-zero state at level $h + v$.

Algorithm 2.6   Viterbi algorithm

By using the Viterbi algorithm we find the number of the best codeword which, together
with cell numbers, uniquely determines the approximating vector y.

Trellises were introduced by Forney in 1974 (Forney 1974) as a means of describing
the Viterbi algorithm for decoding convolutional codes. Later on Bahl *et al.* (Bahl *et al.*
1974) showed that block codes can also be described by trellises.

The considered lattice quantizer is a variable-rate type quantizer. As we saw, it is
based on infinite lattices and uses two variable-rate uniform scalar quantizers. One more
class of lattice quantizer which appeared historically before the considered one is based
on n-dimensional lattice-type trellis codes (Calderbank and Sloane 1987). These lattice
quantizers are suitable for fixed-rate applications.

The trellis encoder maps a sequence of binary symbols into a sequence of points
belonging to an n-dimensional lattice. In fact, instead of using the full lattice the encoder
selects points from a set Λ which is a subset of the lattice restricted by an interval
$[-A, A]$ say, in each dimension. Assume that the interval $[-A, A]$ contains $Q = 2^q$
lattice points. Then in total the number of points in Λ is equal to 2^{qn}. By applying the
set partitioning technique from (Ungerboeck 1982) Λ is divided into 2^N subsets Λ_i in
such a way that Λ is a disjoint union of 2^N subsets

$$\Lambda = \bigcup_{i=1}^{2^N} \Lambda_i.$$

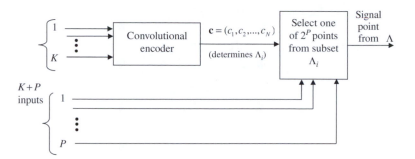

Figure 2.11   Channel trellis encoder

The size of each subset is equal to 2^{qn-N} and parameters (q, n, N) depend on the required quantization rate.

At each time instant in each of 2^{ν} states the encoder of a rate K/N binary convolutional code with constraint length ν selects one of 2^K allowable subsets out of 2^N subsets Λ_i. The 2^K allowable subsets are determined by the K input bits and the current encoder state.

Similar to the earlier considered variable-rate lattice quantizer this quantizer uses the properly modified decoding procedure developed for error-correcting codes. For this reason we start the description of the quantization procedure with consideration of the corresponding channel trellis encoder.

The channel trellis encoder from Ungerboeck (1982) is shown in Fig. 2.11. The input binary sequence is split into blocks of size $K + P$. The first K bits from each block enter the convolutional encoder. The N output bits of the convolutional encoder determine a subset Λ_i where i is an integer whose binary representation is equal to the code branch $\mathbf{c}$. The remaining P "uncoded" bits determine one of the 2^P points in the chosen subset, i.e. $P = qn - N$. Typically, for channel coding we use 1D and 2D integer lattices. The values $n = 1$ and $n = 2$ correspond to the amplitude and quadrature amplitude modulation, respectively.

The quantizer based on such a code for the case when Λ is a subset of Z^n is considered in (Marcellin and Fischer 1990). Quantization is basically equivalent to decoding and is performed in two steps. First, at each level of the trellis we find the points closest to the sample of input sequence in all subsets and store their numbers and the corresponding quantization errors. In such a way we obtain branch metrics and in Ungerboeck's terminology "uncoded bits." Second, by using the Viterbi decoding algorithm we find the best approximating sequence of subsets which is computed as the shortest path in the trellis (path with the minimal sum of branch metrics). The output of the quantizer is two sequences: a sequence of subset numbers and a sequence of the numbers of the closest points in these subsets.

Figure 2.12 illustrates one level of the trellis labeled by the numbers of subsets of approximating values D_0, D_1, D_2, D_3 and the partition of the approximating set into subsets. For this quantizer $q = 3, n = 1$, and $N = 2$. Strictly speaking, in this case the union of the subsets is not a lattice but represents a subset of normalized lattice points. It is easy to see that to point out the number of the subset we spend one bit (symbol 0 or 1 in Fig. 2.12). To select the point inside the subset, one more bit is required since each

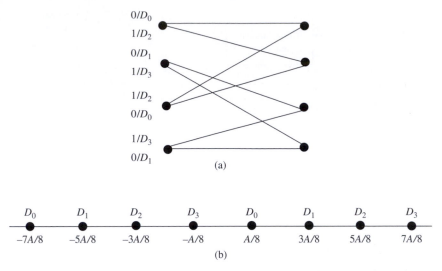

Figure 2.12   (a) Trellis labeled by subset numbers. (b) Partition of the approximating set into subsets

subset contains two approximating values. Thus we obtain that the quantization rate is equal to 2 bits/sample. To complete the description of the quantizer it is necessary to specify the normalizing coefficient A which depends on the energy of the input signal.

Problems

2.1   Assume that the bandlimited signal $x(t) = \cos 4\pi t$ is sampled at f_s samples/s. Find the Fourier transform of the sampled signal and the output of the ideal lowpass filter with cutoff frequency $f_s/2$. Compare the output of the ideal lowpass filter with the original signal $x(t)$ ($f_s = 3\,\text{Hz}$; $f_s = 4\,\text{Hz}$; $f_s = 5\,\text{Hz}$).

2.2   Assume that the bandlimited signal

$$x(t) = \frac{AT\,\sin(\pi t/T)}{\pi t}$$

is sampled at f_s samples/s. Find the Fourier transform of the sampled signal. Find the output of the ideal lowpass filter with cutoff frequency $f_s/2$. Compare the output of the ideal lowpass filter with the original signal $x(t)$ ($f_s \geq 1/T$; $1/(2T) < f_s < 1/T$).

2.3   Consider the rectangular impulse

$$x(t) = \begin{cases} A, & \text{if } -T/2 \leq t \leq T/2 \\ 0, & \text{otherwise.} \end{cases}$$

Find the Fourier transform of the signal $x(t)$. Compute the value f_H of the effective bandwidth with respect to $\theta = 0.9$.

2.4   Consider a source with bounded output (the minimal value is -3, the maximal value is $+3$) that generates the following sequence:

−1.45, 0.41, −0.89, 1.57, −0.64, 0.68, 0.37, −0.76, −2.01, 0.10, −0.85, 0.77, 0.66, 1.84, 0.74, −0.49, 0.54, −0.86, 0.14, 0.11

which enters the quantizer.

Find the number of quantization cells, the approximating values, and the thresholds for the uniform scalar quantizer of rate 2 bits/sample. Write down the quantized and the reconstructed sequences. Estimate the mean squared quantization error and the number of bits for storing an input sequence.

2.5 Quantize the sequence

−2.3, −1.78, 1.40, −0.10, 0.76, 0.65, 0.44, −0.58, −0.53, 0.05, −0.77, −0.63, 1.89, 0.48, −0.80, 0.40, −0.71, −0.93, 0.16, 2.4

which is generated by a source with bounded output (the minimal value is −3, the maximal value is +3) by the uniform fixed-rate scalar quantizer of rate 3; 2; 1 bits/sample, respectively. Compute the mean squared quantization error. Estimate the probabilities of the approximating values. Estimate the entropy of the discrete source formed by the quantizer outputs. Compute the number of bits for storing the quantized sequence under the assumption that the quantizer is followed by a variable-length coder and the probability distribution is known to the decoder.

2.6 For the input sequence given in Problem 2.4, construct the optimal nonuniform scalar quantizer of rate 2 bits/sample by using the Lloyd–Max procedure.

2.7 Assume that X is a Gaussian variable with zero average and variance equal to 1. For a fixed-rate scalar quantizer of rate 1 bit/sample, assume that the given set of thresholds is $\{t_0 = -\infty,\ t_1 = 0,\ t_2 = \infty\}$ and the given set of approximating values is $\{y_1 = -0.8,\ y_2 = 0.8\}$. Compute the probabilities of the approximating values. Estimate the entropy of the discrete source formed by the quantizer outputs. Estimate the mean squared quantization error. Estimate the number of bits for storing an input sequence.

2.8 For the random Gaussian variable given in Problem 2.7, find the thresholds and approximating values for the nonuniform optimal (Lloyd–Max) scalar quantizer of rate 2 bits/sample. Compute the MSE.

2.9 Assume that for vector quantization the following codebook is used:

$$
\begin{aligned}
y_1 &= (-0.04,\ -0.03,\ 0.85,\ 0.32) \\
y_2 &= (0.41,\ -0.35,\ 0.05,\ 0.42) \\
y_3 &= (-0.07,\ 0.38,\ 1.05,\ 0.02) \\
y_4 &= (0.72,\ -0.06,\ 0.26,\ -0.29)
\end{aligned}
$$

and that the input of the quantizer is the following sequence:

−1.47, 1.33, −0.21, 0.08, 1.57, 0.87, 1.02, 0.04, 0.55, 0.11, 1.22, −0.76, 2.05, 0.41, 0.53, −0.19, 0.81, 0.12, 1.11, −0.11.

Write down the quantized sequence and the reconstructed sequence. Estimate the mean squared quantization error and the number of bits for storing an input sequence.

3 Elements of rate-distortion theory

Rate-distortion theory is the part of information theory which studies data compression with a fidelity criterion. In this chapter we consider the notion of rate-distortion function which is a theoretical limit for quantizer performances. The Blahut algorithm for finding the rate-distortion function numerically is given. In order to compare the performances of different quantizers, some results of the high-resolution quantization theory are discussed. Comparison of quantization procedures for the source with the generalized Gaussian distribution is performed.

3.1 Rate-distortion function

Each quantization procedure is characterized by the average distortion D and by the quantization rate R. The goal of compression system design is to optimize the rate-distortion tradeoff. In order to compare different quantizers, the *rate-distortion function* $R(D)$ (Cover and Thomas 1971) is introduced. Our goal is to find the best quantization procedure for a given source. We say that for a given source at a given distortion $D = D_0$, a quantization procedure with rate-distortion function $R_1(D)$ is better than another quantization procedure with rate-distortion function $R_2(D)$ if $R_1(D_0) \leq R_2(D_0)$. Unfortunately, very often it is difficult to point out the best quantization procedure. The reason is that the best quantizer can have very high computational complexity or sometimes, even, it can be unknown. On the other hand, it is possible to find the best rate-distortion function without finding the best quantization procedure. This theoretical lower limit for the rate at a given distortion is provided by the *information rate-distortion function* (Cover and Thomas 1971). We will denote it as $H(D)$ (analogously to the notion of entropy function in the theory of lossless coding). The *information rate-distortion function* does not depend on a specific coding method and is determined only by the properties of a given source. Before giving a definition of $H(D)$, we define other quantities of interest.

1. Let X be a continuous random variable with probability density function (pdf) $f(x)$ and Y be an approximating set with pdf $f(y)$; then the *mutual information* between X and Y is

$$I(X, Y) = \int_X \int_Y f(x) f(y|x) \log_2 \frac{f(y|x)}{f(y)} \, dx \, dy$$

where $f(y|x)$ is the conditional pdf over the set Y given x. The mutual information represents the average amount of information that the knowledge of the value assumed by the quantized output Y supplies about the original value assumed by X, or vice versa (Berger 1971).

2. Let a discrete-time continuous stationary source generate a sequence $x_1, x_2, \ldots, x_i \in X \subseteq \mathcal{R}$. Assume that Y is an approximating set. For any integer $n = 1, 2, \ldots$ a set X^n of vectors $x = (x_1, x_2, \ldots, x_n)$ is determined. Let Y^n be an approximating set of vectors $y = (y_1, y_2, \ldots, y_n)$, $y_i \in Y$. Then the *mutual information* between X^n and Y^n is determined as

$$I(X^n, Y^n) = \int_{X^n} \int_{Y^n} f(x) f(y|x) \log \frac{f(y|x)}{f(y)} \, dx \, dy$$

where $f(x)$ and $f(y)$ are the pdfs over set X^n and Y^n, respectively, $f(y|x)$, $x \in X^n$, denotes a conditional pdf over Y^n given x.

It is convenient to suppose that $X^n = (X_1, X_2, \ldots, X_n)$, where X_i is a random variable generated at the ith time moment. Analogously, $Y^n = (Y_1, Y_2, \ldots, Y_n)$. Notice that, in general, the pdf $f_i(x_i, y_i)$ determining the probability distribution over the set $X_i Y_i$ depends on i.

3. A source is *memoryless* if for any n and any $x \in X^n$, $x = (x_1, x_2, \ldots, x_n)$,

$$f(x) = \prod_{i=1}^{n} f(x_i)$$

where $f(x_i) = f(x)$ for all i, $x \in X$.

4. The *differential entropy* $h(X)$ of a continuous random variable X with a pdf $f(x)$ is defined as

$$h(X) = -\int_X f(x) \log_2 f(x) \, dx$$

and the *conditional differential entropy* of a continuous random variable X given the continuous random variable Y is defined as

$$h(X|Y) = -\int_X \int_Y f(x, y) \log_2 f(x|y) \, dx \, dy$$

where $f(x, y)$ is a joint pdf of X and Y.

5. The mutual information $I(X, Y)$, where X and Y are continuous, is determined by

$$I(X, Y) = h(X) - h(X|Y) = h(Y) - h(Y|X),$$

if X and Y are discrete then $I(X, Y)$ is determined via unconditional and conditional entropies of X and Y, respectively,

$$I(X, Y) = H(X) - H(X|Y) = H(Y) - H(Y|X).$$

The mutual information $I(X, Y)$ ($I(X^n, Y^n)$) expresses the average information that Y (Y^n) contains about X (X^n) or, from a source coding point of view, the average information that is required to reconstruct X (X^n) with a given accuracy. It is intuitively clear that it should be minimized over all quantizer mappings. We are only interested in such quantizer mappings which result in an average distortion less than or equal to a given value, D, say. Now we are ready to introduce the definition of the *information rate-distortion function*.

- For the discrete-time continuous stationary source given in definition (2), let $d(x, y)$ be a fidelity criterion, then the *information rate-distortion function of the continuous stationary source with respect to this fidelity criterion* is

$$H(D) = \inf_n \{H_n(D)\}$$

where

$$H_n(D) = \min_{\{f(y|x):D_n \leq D\}} \left\{ \frac{1}{n} I(X^n, Y^n) \right\}$$

where

$$D_n = \frac{1}{n} \sum_{i=1}^{n} \int_{X_i Y_i} d(x_i, y_i) f(x_i) f_i(y_i|x_i) \, dx_i \, dy_i$$

denotes the average quantization error.

It can be shown that for the memoryless source given in definition (3)

$$H(D) = \min_{\{f(y|x): \int_X \int_Y d(x,y)f(x)f(y|x)dx \leq D\}} \{I(X, Y)\}$$

where $d(x, y)$ denotes a fidelity criterion.

Notice that in definitions of the mutual information sets X and (or) Y can be discrete. Then we should replace the corresponding integrals in the definitions by sums and the corresponding pdfs by probabilities. In the definition of the information rate-distortion function the type of approximating set (continuous or discrete) is determined by the distortion measure which can be continuous or discrete. For the squared distortion measure, even if X is discrete, the approximating set is assumed to be continuous and the above definitions are valid in the present form, where $f(x)$ is a generalized pdf expressed via the Dirac delta function.

The inverse of $H(D)$ is called the *distortion-rate function* and for the memoryless source it is defined as

$$D(R^*) = \min_{\{f(y|x):I(X,Y) \leq R^*\}} \left\{ \int_X \int_Y d(x, y) f(x) f(y|x) dx \right\}. \tag{3.1}$$

The function $D(R^*)$ is a lower bound on the average distortion for a given maximum average rate R^*.

When dealing with the squared fidelity criterion we say about the maximum quantization gain $g_{max}(R)$ for the given source with the variance D^*. It is determined as

$$g_{max}(R) = 10 \log_{10} \frac{D^*}{D(R)}$$

and measured in decibels (dBs). In rate-distortion theory $H(D)$ plays the same role as the entropy $H(X)$ (for symbol-by-symbol lossless coding of discrete sources) if the source samples are independent or as the entropy rate $H(X/X^\infty)$ (for block lossless coding of discrete sources) if the source samples form a stationary process. As was mentioned before, the main difference with respect to lossless coding theory is that now we are interested in the minimum possible average rate under condition that an average distortion is less than or equal to a given value.

In the sequel we restrict ourselves to consider only discrete-time continuous sources with continuous distortion measures, although $H(D)$ can be determined for the discrete sources as well. Below, we formulate without proof the rate-distortion theorem and its converse for a discrete-time stationary random process.

Theorem 3.1 *For a discrete-time stationary and ergodic continuous source with rate-distortion function $H(D)$ with respect to the squared error $d(x, y) = (x - y)^2$, there exists an n_0 such that for all $n > n_0$ and for any $\delta_1 > 0$ and $\delta_2 > 0$ there exists an (R, D_n)-code of code length n with code rate $R \leq H(D) + \delta_1$ for which the mean squared error D_n is less than or equal to $D + \delta_2$. In other words, when the distortion is arbitrarily close to D there exists a code with code rate arbitrarily close to $H(D)$.*

Theorem 3.2 *(Converse of Theorem 3.1.) For the source given in Theorem 3.1 there does not exist a code for which simultaneously the mean squared error would be less than or equal to D and the rate R would be less than $H(D)$.*

In order to compare quantizers, we use $H(D)$ as a benchmark of the quantizer performance. However, the problem is that $H(D)$ can be found as a closed form solution only for a few simple sources.

Consider a Gaussian memoryless source with squared error distortion. For such a source $H(D)$ is determined only by a one-dimensional (1D) pdf. The following theorem holds:

Theorem 3.3 *The rate-distortion function $H(D)$ for an $N(0, \sigma^2)$ source with squared error distortion is*

$$H(D) = \begin{cases} \frac{1}{2} \log_2(\sigma^2/D), & \text{if } D \leq \sigma^2 \\ 0, & \text{if } D > \sigma^2 \end{cases} \tag{3.2}$$

where $N(0, \sigma^2)$ denotes the Gaussian variable with mathematical expectation zero and variance σ^2.

The plot of $H(D)$ for an $N(0, 1)$ source is presented in Fig. 3.1.
The corresponding distortion-rate function is

$$D(R) = \sigma^2 2^{-2R}. \tag{3.3}$$

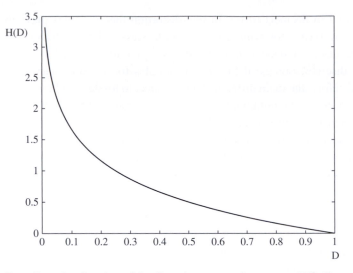

    Rate-distortion function of the Gaussian memoryless source $N(0, 1)$

The maximum quantization gain $g_{max}(R)$ is

$$g_{max}(R) = 10\log_{10}\frac{\sigma^2}{\sigma^2 2^{-2R}} = 10\log_{10} 2^{2R}. \tag{3.4}$$

From (3.4) follows the so-called "6 dB Rule" which is often used for telephone Pulse Code Modulation (PCM). It means that increasing the quantization rate by 1 bit leads to a reduction of quantization noise energy by 6 dB. According to this rule, the maximum amplitude input PCM provides signal-to-noise ratio $6n + c_0$, where n is the number of bits per sample and c_0 is a constant.

Theorem 3.4   *The rate-distortion function $H(D)$ for the source generating a random variable with the Laplacian distribution*

$$f(x) = (\alpha/2)\exp^{-\alpha|x|}$$

where $-\infty < x < \infty$ with $d(x, y) = |x - y|$ is equal to

$$H(D) = \begin{cases} -\log_2(\alpha D), & \text{if } 0 \leq D \leq 1/\alpha \\ 0, & \text{if } D > 1/\alpha. \end{cases} \tag{3.5}$$

The corresponding distortion-rate function is

$$D(R) = \frac{2^{-R}}{\alpha}.$$

Theorem 3.5   *The rate-distortion function $H(D)$ for the source generating a random variable with uniform distribution*

$$f(x) = \begin{cases} \frac{1}{2A}, & \text{if } |x| \leq A \\ 0, & \text{if } |x| \geq A \end{cases} \tag{3.6}$$

with $d(x, y) = |x - y|$ is equal to

$$H(D) = \begin{cases} -\log_2 \left(\dfrac{1 - \left(1 - \frac{2D}{A}\right)^{1/2}}{2 - \left(1 - \frac{2D}{A}\right)^{1/2}} \right), & \text{if } 0 \le D \le \frac{A}{2} \\ 0, & \text{if } D \ge \frac{A}{2}. \end{cases} \tag{3.7}$$

Consider a source, which generates a discrete-time *stationary random Gaussian process*. In other words, let for any $n = 1, 2, \ldots$ a random vector $X = (X_1, X_2, \ldots, X_n)$ at the output of the source be the Gaussian vector with covariance matrix Λ_n and vector of average values $m = (m_1, m_2, \ldots, m_n)$. Its pdf has the form

$$f_n(x) = \frac{1}{(2\pi)^{n/2} |\Lambda_n|^{1/2}} \exp\left\{ -\frac{1}{2} (x - m) \Lambda_n^{-1} (x - m)^{\mathrm{T}} \right\}.$$

The property of stationarity means that the n-dimensional pdfs, $n = 1, 2, \ldots$ of the source vectors do not depend on the shift of the time axis. In other words, the pdfs of the vectors $X = (X_1, X_2, \ldots, X_n)$ and $X_j = (X_{j+1}, X_{j+2}, \ldots, X_{j+n})$ are identical.

It follows from the property of stationarity that $m_i = m_j = m$ and each entry

$$\lambda_{ij} = \mathrm{E}\{(X_i - m)(X_j - m)\}$$

of the covariance matrix Λ_n depends only on the absolute value of the difference $|i - j|$, that is,

$$\lambda_{ij} = \lambda_{ji} = \lambda_{|i-j|}. \tag{3.8}$$

The matrices with elements satisfying (3.8) are called *Toeplitz's matrices*. Let $\tau = i - j$, $\tau = -(n - 1), \ldots, 0, 1, \ldots, n - 1$. It follows from (3.8) that

$$\lambda_\tau = \lambda_{-\tau}$$

and that the covariance matrix of the segment of the stationary process is fully determined by the n numbers λ_τ, $\tau = 0, 1, \ldots, n - 1$, representing the covariance moments of the random variables X_i and X_j spaced τ samples apart.

Assume for simplicity that $m = 0$ and that $\lim_{\tau \to \infty} \lambda_\tau = 0$. Consider the following Fourier series expansion

$$\sum_{\tau=-\infty}^{\infty} \lambda_\tau e^{-j2\pi f\tau} = N(f), \; -1/2 \le f \le 1/2.$$

The function $N(f)$ is called *power spectral density* of the random process generated by a stationary source. A function is bounded if the set of its values is bounded and any set of real or complex numbers is bounded if the set of absolute values of these numbers has an upper bound. If $N(f)$ is a continuous and bounded frequency function (we will consider only this case), then there exists the inverse transform, which allows us to express the covariance moments λ_τ, $\tau = 0, 1, \ldots$, via the power spectral density $N(f)$:

$$\lambda_\tau = \int_{-1/2}^{1/2} N(f) e^{j2\pi f \tau} \, df.$$

The main properties of the power spectral density are the following:

- $N(f)$ is a real function.
- $N(f) = 2 \sum_{\tau=1}^{\infty} \lambda_\tau \cos(2\pi f \tau) + \lambda_0$.
- It follows from the previous property that $N(f)$ is an even function, that is, $N(f) = N(-f)$, and thereby

$$\lambda_\tau = \int_{-1/2}^{1/2} N(f) \cos(2\pi f \tau) \, df.$$

- $N(f) \geq 0$ for all $f \in \left[-1/2, 1/2\right]$.
- $\lambda_0 = \int_{-1/2}^{1/2} N(f) \, df$ is the variance of the random process with power spectral density $N(f)$.

The power spectral density characterizes how the power of the process is distributed over the frequencies.

Theorem 3.6 *The rate-distortion function $H(D)$ with squared error distortion for the discrete-time stationary random Gaussian process with bounded and integrable power spectral density $N(f)$ is computed as follows:*

$$H(D) = \frac{1}{2} \int_{-1/2}^{1/2} \log_2 \left(\max \left\{ 1, \frac{N(f)}{\theta} \right\} \right) df, \tag{3.9}$$

$$\int_{-1/2}^{1/2} \min \{\theta, N(f)\} \, df = D. \tag{3.10}$$

The parameter θ is determined by (3.10) which can be interpreted as "water-filling" (Cover and Thomas 1971) as shown in Fig. 3.2. The hatched area can be interpreted as the frequency distribution of the error process called *reconstruction error spectral density*. The square of the hatched area is equal to D. The areas under $N(f)$, where $N(f) > \theta$, correspond to the *preserved spectral density* of the quantizer output. In order to find $H(D)$ it is necessary to integrate the function $\log_2 (N(f)/\theta)$ over those frequency areas, where $N(f) \geq \theta$ (in our example we integrate over the intervals $(-f_1, -f_2)$, $(-f_3, f_3)$, (f_2, f_1)).

If the samples of the discrete-time stationary random Gaussian process are independent Gaussian variables with the same variance equal to σ^2, then (3.10) reduces to (3.2). In other words, in this case the curve $H(D)$ coincides with the curve shown in Fig. 3.1. If the samples of the Gaussian process are dependent, then $H(D)$ will lie below the curve

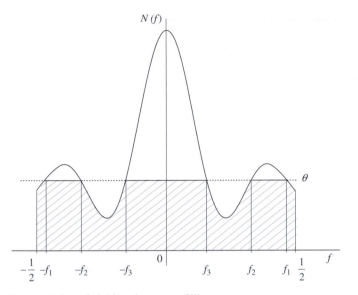

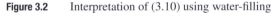

Figure 3.2   Interpretation of (3.10) using water-filling

shown in Fig. 3.1. A first-order Gaussian Markov stationary process has the covariance matrix in the form

$$\Lambda_n = \sigma^2 \begin{pmatrix} 1 & \rho & \rho^2 & \ldots & \rho^{n-1} \\ \rho & 1 & \rho & \ldots & \rho^{n-2} \\ \rho^2 & \rho & 1 & \ldots & \rho^{n-3} \\ \ldots & \ldots & \ldots & \ldots & \ldots \\ \rho^{n-1} & \rho^{n-2} & \rho^{n-3} & \ldots & 1 \end{pmatrix}$$

where $\rho = \mathrm{E}\{(X_i - \mathrm{E}\{X_i\})(X_{i+1} - \mathrm{E}\{X_{i+1}\})\}$ is the correlation coefficient between two neighboring samples and σ^2 denotes the variance of the samples. In Fig. 3.3 the rate-distortion functions $H(D)$ for the Gaussian Markov process with $\rho = 0$, $\rho = 0.5$, and $\rho = 0.9$ are shown.

In general, the function $H(D)$ has the following properties:

- It is a nonincreasing and convex downwards function of D.
- There exists some value D_0 such that $H(D) = 0$ for all $D \geq D_0$.
- Let X be a random variable, c be a constant, and $H(D)$ be a rate-distortion function of X. Then the rate-distortion function of the random variable $X + c$ is also equal to $H(D)$.

3.2   The Blahut algorithm

In this section we consider the Blahut algorithm (Blahut 1972), which allows us to find a numerical value of $H(D)$ for any distortion. Notice that the Blahut algorithm finds an approximation of $H(D)$ under the assumption that X and Y are discrete sets. Therefore,

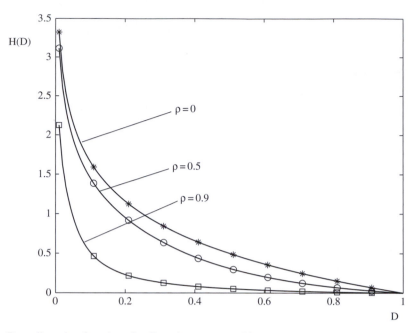

Figure 3.3   Rate-distortion functions for Gaussian sources with memory

in further derivations we deal only with discrete random variables. We introduce some definitions.

- The *relative entropy (or Kullback Leibler distance)* (Cover and Thomas 1971) between two probability mass functions $p(x)$ and $q(x)$ is defined by

$$L(p\|q) = \sum_{x \in X} p(x) \log_2 \frac{p(x)}{q(x)}.$$

The *relative entropy* is often referred to as a distance between two distributions $p(x)$ and $q(x)$. Formally it is not a distance since it does not satisfy the triangle inequality and is not symmetric. It represents the average of the logarithm of the maximum-likelihood ratio of the two pdfs and can be considered as a measure of ineffectiveness when we substitute $p(x)$ by $q(x)$. For example, if we know the true probability distribution $p(x)$, we can construct a code with average codeword length $H(p(x))$. ($H(p(x))$ denotes entropy $H(X)$ for the random variable X with the probability mass function $p(x)$.) If we instead construct a code for the distribution $q(x)$, then the average codeword length will be equal to $H(p(x)) + L(p\|q)$ bits.

One more definition of the mutual information $I(X, Y)$ can be given via the relative entropy:

- Consider two random variables X and Y with probability mass functions $p(x)$ and $p(y)$, respectively. Let $p(x, y)$ be the joint probability mass function. Then the *mutual information $I(X, Y)$* is the *relative entropy* between the joint distribution and the product distribution $p(x)p(y)$, that is,

$$I(X, Y) = \sum_{x \in X} \sum_{y \in Y} p(x, y) \log_2 \frac{p(x, y)}{p(x)p(y)}.$$

The problem of finding $H(D)$ reduces to the problem of minimization of $I(X, Y)$ on condition that $E\{d(x, y)\} = E\{(x - y)^2\} \leq D$. The standard way of solving this problem is to minimize the corresponding Lagrangian function, that is,

$$H(D) = \min_{p(y|x)} \{I(X, Y) - \lambda(E\{d(x, y)\} - D)\}$$
$$= \lambda D + \min_{p(y|x)} \{F(p(x), p(y|x), d(x, y), \lambda)\}$$

where $F(p(x), p(y|x), d(x, y), \lambda) = I(X, Y) - \lambda E\{d(x, y)\}$. In the sequel of this section we will use the notation $I(p(x), p(y|x))$ for $I(X, Y)$.

The essence of the Blahut algorithm can be explained as follows (Cover and Thomas 1971). Assume that we would like to find the minimum Euclidean distance between two convex sets A and B in $\mathcal{R}^n$. The intuitive algorithm consists of the following steps:

Input:
Convex sets $A, B \subseteq \mathcal{R}^n$
Output:
The minimum Euclidean distance between A and B.

repeat

- Take any point $a \in A$ and find $b \in B$ that is closest to a.
- Fix the point b and find a closest point to it in A.

until *some stop condition is fulfilled*

Algorithm 3.1   Idea of the Blahut algorithm

It is clear that at each step of this algorithm the distance decreases. It was shown in (Csiszar and Tusnady 1984) that if the sets are sets of probability distributions and the distance measure is the relative entropy, then the algorithm converges to the minimum relative entropy between the two sets of distributions.

Since we are interested in the conditional minimum of $I(X, Y)$, then in the Blahut algorithm the same idea is applied to find the minimum of the function $F(p(x), p(y|x), d(x, y), \lambda)$. According to the Blahut algorithm, the minimization problem is split into two minimization problems which can be solved analytically.

Let us rewrite $F(p(x), p(y|x), d(x, y), \lambda)$ in the form

$$F(p(x), p(y|x), d(x, y), \lambda) = \sum_{x \in X} \sum_{y \in Y} p(x)p(y|x) \log_2 \frac{p(y|x)}{p(y)}$$
$$- \lambda \sum_{x \in X} \sum_{y \in Y} p(x)p(y|x)d(x, y). \qquad (3.11)$$

Blahut showed that the problem of minimizing $F(\cdot)$ can be reduced to the double minimization of the following function

$$\min_{r(y)} \min_{p(y|x)} \left\{ \sum_{x \in X} \sum_{y \in Y} p(x)p(y|x) \log_2 \frac{p(y|x)}{r(y)} - \lambda \sum_{x \in X} \sum_{y \in Y} p(x)p(y|x)d(x, y) \right\}.$$

(3.12)

It is easy to see that the difference between (3.12) and (3.11) is that we fixed $p(y) = r(y)$ and minimize $F(\cdot)$ separately over $r(y)$ and $p(y|x)$. Finally, we obtain

$$H(D) = \lambda D$$

$$+ \min_{r(y)} \min_{p(y|x)} \left\{ \sum_{x \in X} \sum_{y \in Y} p(x)p(y|x) \log_2 \frac{p(y|x)}{r(y)} - \lambda \sum_{x \in X} \sum_{y \in Y} p(x)p(y|x)d(x, y) \right\}$$

where

$$D = \sum_{x \in X} \sum_{y \in Y} p(x)p^*(y|x)d(x, y),$$

and $p^*(y|x)$ achieves the above minimum. It was also shown in (Blahut 1972) that if $p(y|x)$ is fixed, then the optimal value of $r(y)$ is determined as

$$r(y) = \sum_{x \in X} p(x)p(y|x),$$

that is, it coincides with $p(y)$. If $r(y)$ is fixed, then the optimal value of $p(y|x)$ is equal to

$$p(y|x) = \frac{r(y) \exp(\lambda d(x, y))}{\sum_{y \in Y} r(y) \exp(\lambda d(x, y))}.$$

In order to form a stopping rule, the following bounds on $H(D)$ are used:

$$H(D) \leq \lambda D - \sum_{x \in X} p(x) \log_2 \sum_{y \in Y} \exp(\lambda d(x, y))r(y) - \sum_{y \in Y} r(y)c(y) \log_2 c(y),$$

$$H(D) \geq \lambda D - \sum_{x \in X} p(x) \log_2 \sum_{y \in Y} \exp(\lambda d(x, y))r(y) - \max_{y \in Y} \left\{ \log_2 c(y) \right\}$$

where

$$c(y) = \sum_{x \in X} p(x) \frac{\exp(\lambda d(x, y))}{\sum_{y \in Y} r(y) \exp(\lambda d(x, y))}.$$

Below, we give a description of the Blahut algorithm:

Input:
probability distribution $p(x)$
parameter λ
a distortion measure $d(x, y)$
a convergence threshold $\epsilon > 0$
a number of iterations N_{I}
Output:
The average distortion D and the corresponding rate $H(D)$

Initialization: Choose an initial distribution $r(y)$
Set the current number of iterations $N = 0$

repeat
 Compute the probability scaling factor $c(y)$ and update the probability distribution $r(y)$ and the thresholds T_{U} and T_{L}:

$$c(y) = \sum_{x \in X} p(x) \frac{\exp(\lambda d(x, y))}{\sum_{y \in Y} r(y) \exp(\lambda d(x, y))}$$

$$r(y) = r(y)c(y)$$

$$T_{\mathrm{U}} = -\sum_{y \in Y} r(y) \log_2(c(y))$$

$$T_{\mathrm{L}} = -\max_{y \in Y} \log_2(c(y)).$$

Increment the current number of iterations

$$N \leftarrow N + 1$$

until $T_{\mathrm{U}} - T_{\mathrm{L}} < \epsilon$ **or** *number of iterations is greater than* N_{I}
With the conditional probability distribution

$$p(y|x) = \frac{r(y) \exp(\lambda d(x, y))}{\sum_{y \in Y} r(y) \exp(\lambda d(x, y))},$$

compute the average distortion D and the rate $H(D)$ as

$$D = \sum_{x \in X} \sum_{y \in Y} p(x)p(y|x)d(x, y)$$

$$H(D) = \lambda D - \sum_{x \in X} p(x) \log_2 \sum_{y \in Y} \exp(\lambda d(x, y))r(y) - \max_{y \in Y} \log_2 c(y).$$

Algorithm 3.2   The Blahut algorithm

It is convenient to apply the Blahut algorithm to a sequence normalized to the interval [0, 1]. It is performed by adding a bias to the input values and multiplying them by a constant. It is known that adding a bias does not change $H(D)$ of the source. Since we compute the relative distortion, that is, the ratio $D/\text{Var}(X)$, where $\text{Var}(X)$ is the variance of the normalized input sequence, then neither will multiplying input by a constant change $H(D)$.

For a discrete-time continuous source, we can also use the Blahut algorithm to approximate $H(D)$. This can be done by first approximating the source using a high-rate scalar uniform quantizer and then calculating the $H(D)$ curve of the quantized source. Assume that we observe the output sequence $x_1, x_2, \ldots, x_k$ of the discrete-time continuous source. Choose the scalar uniform quantizer with large number of quantization cells, say $N = 20-50$. Then we can estimate the probability mass function of the corresponding memoryless source as

$$p_i = \frac{k_i}{k}, i = 1, 2, \ldots, N$$

where k_i is the number of samples x_i (normalized to the interval [0, 1]) belonging to the ith quantization cell $((i - 1)/N, i/N]$ with the approximating value $(2i - 1)/(2N)$.

The distribution $r(y)$ can be initialized, for example, by the uniform distribution, i.e.

$$r(y) = \frac{1}{N}, y \in \left\{ \frac{1}{2N}, \frac{3}{2N}, \ldots, \frac{2N - 1}{2N} \right\}.$$

Notice that formally we should optimize the parameter λ in order to obtain a given rate value. However, in practice instead of doing this we usually choose a predetermined set of values $\lambda_k, k = 0, 1, \ldots, K - 1$ to compute K points of the curve $H(D)$.

3.3   The Shannon lower bound and high-resolution approximations

Searching for the rate-distortion function $H(D)$ is a rather computationally expensive procedure. In practice rather often the Shannon lower bound on $H(D)$ is used. The form of this bound depends on the distortion measure. For the squared error distortion it has the form

$$H(D) \geq R_{\text{L}}(D) = h(X) - \frac{1}{2} \log_2(2\pi e D). \tag{3.13}$$

If $d(x, y) = |x - y|$, then $R_{\text{L}}(D)$ is determined as

$$H(D) \geq R_{\text{L}}(D) = h(X) - \log_2(2eD).$$

In some cases the Shannon lower bound is tight. For example, for the Gaussian memoryless source $N(0, \sigma^2)$ with squared distortion measure we have $H(D) = R_{\text{L}}(D)$ for all $0 < D < \sigma^2$. The same is true for the memoryless source with the Laplacian distribution and the magnitude distortion measure $d(x, y) = |x - y|$.

In order to compare the performances of different quantizers, consider some results of the so-called high-resolution quantization theory. The essence of this theory is to assume that the accuracy of quantization is very high and the pdf of the input is reasonably smooth. Therefore we can say that the pdf is approximately constant over any quantization cell, since the number of cells is large and the cells are small enough. Under this assumption for the uniform scalar quantizer, we can write

$$P(y_i) \approx \delta f(y_i)$$

$$D = \sum_{i=1}^{M} \int_{t_{i-1}}^{t_i} (x - y_i)^2 f(y_i) \, dx$$

$$\approx \sum_{i=1}^{M} \frac{P(y_i)}{\delta} \int_{y_i - \delta/2}^{y_i + \delta/2} (x - y_i)^2 \, dx = \frac{\delta^2}{12}.$$

Estimate the rate of such a quantizer:

$$R = -\sum_{i=1}^{M} P(y_i) \log_2 P(y_i)$$

$$\approx -\sum_{i=1}^{M} P(y_i) \log_2(\delta f(y_i))$$

$$\approx -\sum_{i=1}^{M} \int_{t_{i-1}}^{t_i} f(x) \log_2(\delta f(x))$$

$$= -\int_X f(x) \log_2 f(x) - \log_2 \delta = h(X) - \log_2 \delta.$$

Taking into account that $D = \delta^2/12$, we obtain

$$R(D) = h(X) - \frac{1}{2} \log_2(12D). \tag{3.14}$$

By comparing (3.14) and (3.13) we find that for a memoryless source with the squared fidelity criterion if the step size $\delta \to 0$, the redundancy of the uniform scalar quantizer is

$$R(D) - R_L(D) \leq \frac{1}{2} \log_2 \frac{\pi e}{6} \approx 0.2546.$$

Therefore, for rather high quantization rates a vector quantization of a memoryless source with the squared distortion measure can win not more than 0.2546 bits per sample compared to uniform scalar quantization (Koshelev 1963; Gish and Pierce 1968). Notice that if the quantization rate is low (for example, less than 1 bit per sample), then the gain of vector quantization compared to scalar quantization can be significant.

It follows from the coding theorem (3.1) that by increasing the dimension n of quantizer we can achieve $H(D)$. On the other hand, the complexity of the vector quantizer grows exponentially with n. The important question arises: how quickly does the efficiency of the vector quantizer increase when n grows? By using a high-resolution approximation for the vector quantization, we can answer this question if small quantization errors are assumed.

Assume that the number of cells is large and all cells S_i have the same size and shape, that is, they are congruent. In other words, consider the high-resolution approximation for the lattice quantizer. Under this assumption, the pdf is constant within each cell S_i, $i = 1, 2, \ldots, M$. Then the probability of the ith cell can be written as

$$P(y_i) \approx f(y_i) \int_{S_i} dx = f(c_i) Vol(S_i)$$

where $Vol(S_i) = \int_{S_i} dx$ is called the volume of the ith cell. Under the same assumption (2.16) can be rewritten as

$$D_n \approx \frac{1}{n} \sum_{i=1}^{M} f(y_i) \int_{S_i} \|x - y_i\|^2. \tag{3.15}$$

Since all cells S_i are congruent, then all integrals in (3.15) are equal. Without loss of generality we can put the approximating vector into the origin and consider only the integral over the cell S containing the origin. Then we obtain

$$D_n \approx \frac{1}{n} \sum_{i=1}^{M} f(y_i) \int_S \|x\|^2 dx$$

$$= \frac{1}{n} \sum_{i=1}^{M} f(y_i) Vol(S_i) \frac{\int_S \|x\|^2 dx}{Vol(S_i)} \approx \frac{1}{n} \sum_{i=1}^{M} P(y_i) \frac{\int_S \|x\|^2 dx}{Vol(S)}$$

$$= \frac{1}{n} \frac{U(S)}{Vol(S)} \tag{3.16}$$

where $U(S) = \int_S \|x\|^2 dx$ is called the *second moment* of the Voronoi region.

With the same assumptions the quantization rate can be rewritten as

$$R = -\frac{1}{n} \sum_{i=1}^{M} P(y_i) \log_2 P(y_i) \approx -\frac{1}{n} \sum_{i=1}^{M} P(y_i) \log_2(Vol(S) f(y_i))$$

$$\approx -\frac{1}{n} \sum_{i=1}^{M} \int_{S_i} f(x) \log_2(Vol(S) f(x)) dx$$

$$= -\frac{1}{n} \left(\int_{X^n} f(x) \log_2 f(x) dx + \log_2 Vol(S) \right)$$

$$= h(X) - \frac{1}{n} \log_2(Vol(S)). \tag{3.17}$$

The last equality in (3.17) follows from the fact that we consider a memoryless source and $\int_{X^n} f(x) \log_2 f(x)\, dx = nh(X)$.

Using the Shannon lower bound (3.13) we obtain

$$R(D) \approx R_{\mathrm L}(D) + \frac{1}{2}\log_2(2\pi e D) - \frac{1}{n}\log_2 Vol(S)$$

$$= R_{\mathrm L}(D) + \frac{1}{2}\log_2 \frac{2\pi e D}{Vol^{2/n}(S)}. \tag{3.18}$$

Inserting (3.16) into (3.18) we obtain

$$R(D) \approx R_{\mathrm L}(D) + \frac{1}{2}\log_2 2\pi e G_n \tag{3.19}$$

where

$$G_n = \frac{1}{n}\frac{\int_S \|x\|^2\, dx}{\left(\int_S dx\right)^{1+2/n}}$$

is called the *normalized second moment* of the Voronoi region (Zador 1982). The following bounds on G_n hold (Conway and Sloane 1988; Zador 1982):

$$\frac{1}{(n+2)\pi}\Gamma\left(\frac{n}{2}+1\right)^{2/n} \le G_n \le \frac{1}{n\pi}\Gamma\left(\frac{n}{2}+1\right)^{2/n}\Gamma\left(1+\frac{2}{n}\right) \tag{3.20}$$

where

$$G_n = \frac{1}{(n+2)\pi}\Gamma\left(\frac{n}{2}+1\right)^{2/n}$$

is the *normalized second moment of the n-dimensional sphere* and $\Gamma(\alpha)$ denotes the so-called Gamma function which is equal to $(\alpha-1)!$ if α is a natural number. It follows from (3.20) that when n grows, the lower and the upper bounds coincide and

$$G_n \to \frac{1}{2\pi e} = 0.058550\ldots \text{ when } n\to\infty.$$

Substituting G_n by the value $1/(2\pi e)$ in (3.19) we obtain that $R(D) \approx R_{\mathrm L}(D)$. It means that the minimal possible quantization rate can be achieved if the Voronoi cell has a spherical form. On the other hand, for any finite n, spheres never cover the space $\mathcal{R}^n$ and therefore cannot be the cells of any real quantizer.

The value G_n is tabulated for many popular lattices, that is, lattice quantizers. For example, for the *hexagonal* lattice A_2 considered in Chapter 2, $G_n = 0.080188$, that is, less than for the 1D integer lattice Z ($G_n = 0.08333\ldots = 1/12$), i.e. for scalar quantization.

The normalized second moment for some lattices is given in Table 3.1. For more details see, for example, (Conway and Sloane 1988; Kudryashov and Yurkov 2007).

Thus, we can conclude that for small quantization errors, when the quantizer dimension tends to infinity, the quantization rate $R(D)$ tends to $H(D)$. The maximal

Table 3.1 G_n of lattices

Dimension	Lattice	G_n
1	Z	0.0833
2	A_2	0.0802
3	A_3^*	0.0785
4	D_4	0.0766
5	D_5^*	0.0756
6	E_6^*	0.0742
7	E_7^*	0.0731
8	E_8	0.0717
12	K_{12}	0.0701
16	Λ_{16}	0.0683
24	Λ_{24}	0.0658
∞	–	0.0585

Lattices based on convolutional codes	
$\nu = 2$	0.0665
$\nu = 5$	0.0634
$\nu = 9$	0.0618
$\nu = \infty$	0.0598

asymptotical shaping gain of the lattice quantizer compared to the scalar uniform quantizer is equal to

$$10 \log_{10} \frac{0.0833}{0.0585} = 1.53 \text{ dB.}$$

Using (3.3) it is easy to check that this value corresponds to a 0.2546 bit gain in the quantization rate.

   Lattice quantizers based on convolutional codes represent a class of vector quantizers with rather low computational complexity determined by the constraint length of underlying convolutional codes. It is easy to see that for short and moderate constraint lengths the second normalized moment of such a quantizer is very close to the theoretical limit.

3.4   Comparison of quantization procedures

High-resolution theory allows us to compare performances of different quantizers under the assumption that the quantization rate is rather high (the quantization error is small). For moderate and low quantization rates the relation between quantizers significantly differs for different sources. Also, we should take into account the fixed-rate limitation if such exists.

   We consider a class of parameterized distributions (Farvardin and Modestino 1984) which is described by the pdf

$$f(x) = \frac{\alpha \gamma (\alpha, \sigma)}{2\Gamma(1/\alpha)} \exp \left\{ -(\gamma(\alpha, \sigma)|x - m|)^\alpha \right\},$$

where m is the mathematical expectation, σ^2 is the variance, $\alpha > 0$ denotes a parameter of distribution describing the exponential rate of decay, and

$$\gamma(\alpha, \sigma) = \sigma^{-1} \left[\frac{\Gamma(3/\alpha)}{\Gamma(1/\alpha)} \right]^{1/2}, \tag{3.21}$$

$$\Gamma(x) = \int_0^\infty e^{-t} t^{x-1} dt,$$

and $\Gamma(n+1) = n!$ if n is natural, $\Gamma(1/2) = \sqrt{\pi}$.

The distribution (3.21) is called the *generalized Gaussian distribution*. For $\alpha = 2$ it corresponds to the Gaussian distribution

$$f(x) = \frac{1}{\sqrt{2\pi}\sigma} \exp \left\{ \frac{-(x-m)^2}{2\sigma^2} \right\}$$

and for $\alpha = 1$ we obtain the Laplacian distribution

$$f(x) = \frac{1}{\sqrt{2}\sigma} \exp \left\{ -\sqrt{2} \frac{|x-m|}{\sigma} \right\}.$$

If α tends to infinity, then distribution (3.21) tends to the uniform distribution. The generalized Gaussian distribution with parameter $\alpha \leq 0.5$ provides a useful model for the 1D probability distribution of transform coefficients for a wide class of discrete transforms (for details, see Chapter 5). The pdfs of the generalized Gaussian distribution with $\alpha = 2.0, 1.0, 0.5$ are shown in Fig. 3.4.

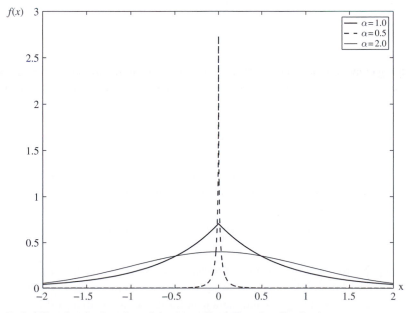

Figure 3.4 Probability density function of the generalized Gaussian distribution

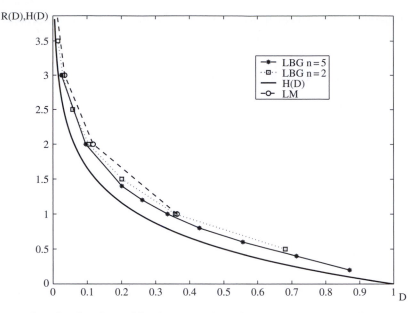

Figure 3.5   Rate-distortion functions of fixed-rate quantizers for the generalized Gaussian memoryless source with $\alpha = 2.0$

Next, we compare the rate-distortion functions of the quantizers described above for the source with the generalized Gaussian distribution where $\alpha = 2.0$ and $\alpha = 0.5$. For simplicity of computation we consider the source with the generalized Gaussian distribution having $m = 0$ and $\sigma^2 = 1$ and assume the squared error distortion measure. We start the comparison with the fixed-rate quantizers.

It is evident that if the quantization rate is *fixed*, then the optimum scalar nonuniform quantizer always has a better rate-distortion function $R(D)$ than the scalar uniform quantizer. Thus, we consider only the scalar nonuniform quantization optimized by the Lloyd–Max (LM) procedure. In Fig. 3.5, for the source with $\alpha = 2.0$ the rate-distortion function of the LM-quantizer is compared with the rate-distortion functions of the vector quantizers with dimensions $n = 2$ and $n = 5$ optimized by the LBG procedure. The function $H(D)$ of the source is plotted in the same figure.

In Fig. 3.6 the curves of the same name are shown for the source with $\alpha = 0.5$.

It follows from Figs 3.5 and 3.6 that when using a nonuniform scalar quantizer with fixed rate, it is impossible to obtain a quantization rate less than 1 bit/sample. The vector quantization reduces the quantization redundancy, that is

$$\max_{D}\{R(D) - H(D)\},$$

compared to scalar quantization. Furthermore, the vector quantizer provides code rates less than 1 bit/sample. It is easy to see that for the source with $\alpha = 2.0$ the redundancy of the quantizers is significantly less than for the source with $\alpha = 0.5$.

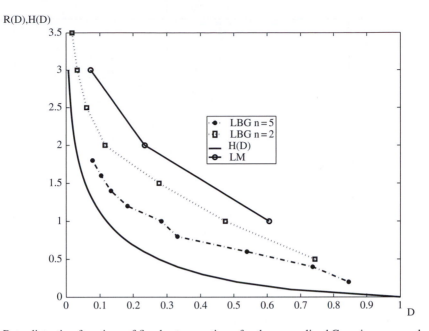

Figure 3.6   Rate-distortion functions of fixed-rate quantizers for the generalized Gaussian memoryless source with $\alpha = 0.5$

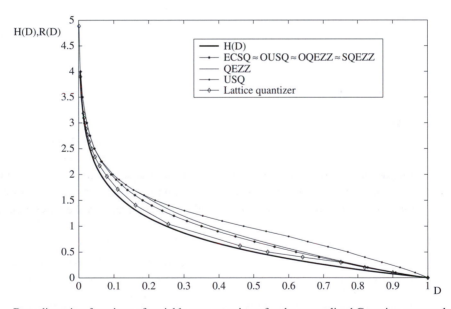

Figure 3.7   Rate-distortion functions of variable-rate quantizers for the generalized Gaussian memoryless source with $\alpha = 2.0$

In Figs 3.7 and 3.8, performances of variable-rate quantizers for the sources with $\alpha = 2.0$ and $\alpha = 0.5$ are illustrated. Notice that the corresponding curves in Fig. 3.8 are given in a logarithmic scale. Some simulation results shown in Fig. 3.8 were presented in (Kudryashov and Porov 2005).

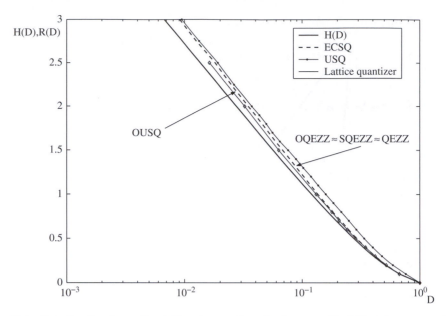

Figure 3.8   Rate-distortion functions of variable-rate quantizers for the generalized Gaussian memoryless source with $\alpha = 0.5$

We conclude that:

- For rates greater than or equal to 2.5 bits/sample, the performances of the Uniform Scalar Quantizer (USQ) and the Optimal Uniform Scalar Quantizer (OUSQ) are almost coinciding with the performances of the entropy constrained optimal scalar quantizer (ECSQ). They are close to the asymptotic estimate $R_L(D) + 0.255$ bits/sample.

- For low rates, the USQ performs worse than the ECSQ.

- For low rates, the Optimum and Suboptimum Quantizers with Extended Zero Zone (OQEZZ, SQEZZ) provide almost the same rate-distortion functions as the ECSQ. For $\alpha = 0.5$ these quantizers at low rates are slightly superior than the OUSQ and for $\alpha = 2.0$ they have the same performances as OUSQ.

- The rate-distortion function of the lattice quantizer, based on the convolutional code with constraint length $\nu = 8$ coincides with the rate-distortion function of the ECSQ for rates less than $0.5 - 0.4$ bits/sample. For higher rates the lattice quantizer wins 0.05 and 0.1 bits per sample compared to the ECSQ for $\alpha = 0.5$ and $\alpha = 2.0$, respectively.

Comparing plots shown in Figs 3.3, 3.5, and 3.7 we see that for sources with memory, all types of scalar quantizer provide rate-distortion functions which are rather far from the achievable rate-distortion function of the given source. Applying vector quantization to sources with memory can reduce a quantization rate compared to scalar quantization but at the cost of an unacceptable increase of computational complexity. This leads to other approaches to efficient representation of multimedia data that can be interpreted as outputs of sources with memory. These approaches are considered in Chapters 6 and 7.

Thus, typically, inputs of multimedia compression systems are samples of multimedia signals pre-quantized by a uniform scalar quantizer. Commonly used digital multimedia formats are described in the next section.

3.5 Characteristics of digital speech, audio, image, and video signals

In this section we present the signal characteristics and the resulting uncompressed bit-rate necessary to support storage and transmission of digital multimedia data with high quality. Table 3.2 containing these data has separate sections for speech and audio signals, images, and video signals, since their characteristics are very different in terms of frequency range, sampling rates, etc.

The standard way to convert all mentioned analog signals into digital forms is first to sample them and then to quantize the samples using a USQ. In the literature the reader can find the term PCM to mean USQ.

Telephone speech is a signal whose frequencies lie in the band 200–3200 Hz. According to the sampling theorem the standard sampling rate is taken as $2 \times 4000 = 8000$ Hz or, equivalently, the sampling period $T_s = 1/8000 = 0.125$ ms. The standard uniform quantizer spends 16 bits per sample. Thus, the bit-rate 128 kb/s is required without any form of coding or compression. Music as a waveform signal has the band 20–20 000 Hz. The standard sampling rate for CD music is 44.1 kHz. To provide CD quality, 16 bits are required to represent each sample. The overall bit-rate for stereo is thus $2 \times 44\,100 \times 16 = 1.411$ Mb/s.

An image is a 2D array of light values. It can be reduced to a 1D waveform by a scanning process. The smallest element of the image is called a *pixel*. Sometimes in image processing a pixel is called a *dot*. The number of pixels per cm in the image determines its *resolution*. A fax represents a kind of image consisting of black and white dots. We can consider it as a digital image where each pixel is represented by a single bit of information and thereby it is not necessary to quantize each pixel value. The standard way to give the resolution of a fax is in Dots Per Inch (DPI), with 1 inch $= 2.5$ cm. Table 3.2 shows the uncompressed size needed for fax. It can be seen that an ordinary fax of a 26 by 17 cm document, scanned at 300 DPI has an uncompressed size equal to 6.36 Mb. When the pixels of an image can take on shades of gray, the image is said to be a *gray-scale* image. In this case a waveform of the intensity along a scanned line is an analog signal. Usually, each pixel intensity is quantized by simple PCM at 8 bits per pixel. The so-called "true-color" computer image standard uses 8 bits for each of the three (Red, Green, and Blue (RGB)) components of a pixel. Color images displayed on a computer screen at VGA resolution require 2.46 Mb and high-resolution XVGA color images require 18.87 Mb for an uncompressed image.

Video is a sequence of still images. The best-known example of video is broadcast television. Table 3.2 shows the necessary bit-rates for several video types. For standard television including the North American National Television Systems Committee (NTSC) standard and the European Phase Alternation Line (PAL) standard, the uncompressed bit-rates are 176.4 Mb/s (NTSC) and 208.3 Mb/s (PAL). Notice that

Table 3.2 Signal parameters and bit-rates

Speech and audio				
Speech/audio type	Frequency band	Sampling rate	Bits/ sample	Uncompressed bit-rate
Speech	200–3 200 Hz	8 kHz	16	128 kb/s
CD audio	20–20 000 Hz	44.1 kHz	16×2	1.41 Mb/s

Still image			
Image type	Pixels per frame	Bits/ pixel	Uncompressed bit-rate
FAX	3120×2040	1	6.36 Mb
VGA	640×480	8	2.46 Mb
XVGA	1024×768	24	18.87 Mb

Video				
Video type	Pixels/ frame	Frame/s	Bits/ pixel	Uncompressed bit-rate
NTSC	700×525	30	16	176.4 Mb/s
PAL	833×625	25	16	208.3 Mb/s
CIF	352×288	15	12	18.2 Mb/s
QCIF	176×144	10	12	3.0 Mb/s
HDTV	1280×720	60	12	622.9 Mb/s
HDTV	1920×1080	30	12	745.7 Mb/s

16 bits/pixel (instead of 24 bits/pixel) is provided by using the so-called 4:2:2 sub-sampling format with two chrominance samples C_u and C_v for every four luminance samples. For video-conferencing and video-phone applications, smaller format pictures with lower frame rates are standard, leading to the Common Intermediate Format (CIF) and Quarter CIF (QCIF) standards which have uncompressed bit-rates of 18.2 and 3.0 Mb/s, respectively. These standards are characterized by 12 bits/pixel that correspond to the so-called 4:1:1 color subsampling format with one chrominance sample C_u (C_v) for every four luminance samples.

By increasing the number of scan lines and increasing the analog bandwidth, a higher resolution than for standard television signals can be obtained. The digital standard for High-Definition Television (HDTV) (in two standard formats) has requirements for an uncompressed bit-rate of between 622.9 Mb/s and 745.7 Mb/s.

Problems

3.1 Show that for the USQ, when the quantization step $\delta \to 0$ the MSE can be approximated by $\delta^2/12$.

3.2 For the Gaussian memoryless source with zero mean and variance equal to 1, compute the rate-distortion function for the nonuniform optimal scalar (LM) quantizer.

Compare the obtained rate-distortion function with the achievable rate-distortion function $H(D)$ for this source.

3.3 For the Gaussian memoryless source with zero mean and variance equal to 1, compute the rate-distortion function for the uniform scalar variable-rate quantizer. Compare the obtained rate-distortion function with $H(D)$ and with the rate-distortion function of the LM quantizer given in Problem 3.2.

3.4 Compute the maximum coding gain $g_{max}(R)$ for the quantizers given in Problems 3.2 and 3.3. Compare the obtained functions for different rate values. Draw a conclusion.

3.5 Show that for the Gaussian memoryless source with zero mean and variance σ^2 (3.9) reduces to (3.2).

3.6 Derive a formula for the second normalized moment for Z^n.

3.7 Show that the pdf of the generalized Gaussian distribution reduces to the pdf of the Laplace distribution, the Gaussian distribution, and the uniform distribution if $\alpha = 1, 2$, and $\alpha \to \infty$, respectively.

3.8 Consider two random Gaussian variables X and Y with zero mean values and variances σ_x^2, σ_y^2, respectively. Assume that

$$E\{XY\} = \sigma_x \sigma_y \rho.$$

Write down the joint pdf for the variables X and Y.

3.9 Find the mutual information $I(X, Y)$ for the variables given in Problem 3.8. Plot $I(X, Y)$ as a function of ρ.

3.10 By using the Blahut algorithm find the approximation for the achievable rate-distortion function $H(D)$ of the Gaussian memoryless source with zero mean and variance one. Assume that the squared distortion measure is used. Compare the curve obtained with the curve determined by (3.2).

4 Scalar quantization with memory

Multimedia data can be considered as data observed at the output of a source with memory. Sometimes we say that speech and images have considerable *redundancy*, meaning the statistical correlation or dependence between the samples of such sources which is referred to as *memory* in information theory literature. Scalar quantization does not exploit this redundancy or memory. As was shown in Chapter 3, scalar quantization for sources with memory provides a rate-distortion function which is rather far from the achievable rate-distortion function $H(D)$ for a given source. Vector quantization could attain better rate-distortion performance but usually at the cost of significantly increasing computational complexity. Another approach leading to a better rate-distortion function and preserving rather low computational complexity combines linear processing with scalar quantization. First, we remove redundancy from the data and then apply scalar quantization to the output of the memoryless source. Outputs of the obtained memoryless source can be vector quantized with lower average distortions but with higher computational complexity. The two most important approaches of this variety are *predictive coding* and *transform coding* (Jayant and Noll 1984). The first approach is mainly used for speech compression and the second approach is applied to image, audio, and video coding. In this chapter, we will consider *predictive coding* systems which use time-domain operations in order to remove redundancy and thereby to reduce the bit-rate for given quantization error levels.

4.1 Discrete-time filters

Helpful introductions to digital filtering can be found in many textbooks. See, for example, (Mitra 2005). Therefore, we only remind ourselves in brief of some basics of digital filtering in so far as we will need them further on.

- A discrete-time *system* transforms an input sequence $x(nT_s)$ into an output sequence $y(nT_s)$. Mathematically, a *system* is defined as a unique transformation or operator that maps an input sequence $x(nT_s)$ into an output sequence $y(nT_s)$. This is denoted as

$$y(nT_s) = T\{x(nT_s)\}.$$

  In order to simplify notations in the sequel, instead of nT_s we often specify only the number n of the sample value. For example, instead of $x(nT_s)$ we write $x(n)$.

Any discrete-time system can be considered as a *discrete-time filter*. Thus, such discrete-time systems as predictive coding and transform coding are, in fact, kinds of discrete-time filtering. We will consider only linear discrete-time systems (filters). A system T is said to be linear if the following condition holds

$$T\{\alpha x_1(n) + \beta x_2(n)\} = \alpha T\{x_1(n)\} + \beta T\{x_2(n)\}$$

where α and β are arbitrary real or complex constants.

We say that a system is *causal* if the output for $n = n_0$ depends only on the inputs for $n \leq n_0$. A system is *stable* if every bounded input produces a bounded output. We say that a sequence $x(n)$ is causal if $x(n) = 0$ for $n < 0$. Below, we briefly discuss how discrete-time filters can be described.

4.1.1 Description by means of recurrent equations

A discrete-time filter of order N is described by the following linear recurrent equation

$$y(n) = \sum_{i=0}^{M} a_i x(n-i) + \sum_{j=1}^{N} b_j y(n-j) \tag{4.1}$$

where a_i, $i = 0, 1, \ldots, M$, and b_j, $j = 1, 2 \ldots, N$, are constants which do not depend on $x(n)$.

Linear discrete-time filters can be split into two classes: time-invariant discrete-time filters (the parameters a_i and b_j do not depend on n) and time-varying discrete-time filters (the parameters a_i and b_j depend on n). A time-invariant discrete-time filter of order N is shown in Fig. 4.1.

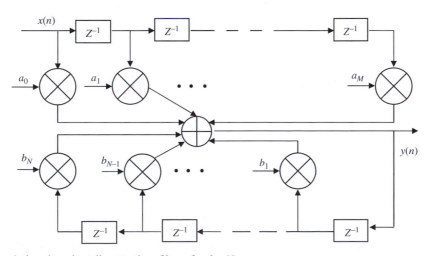

Figure 4.1 A time-invariant discrete-time filter of order N

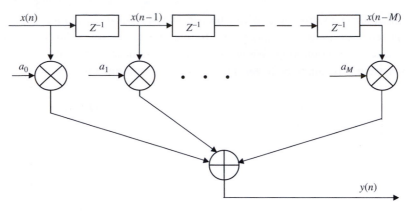

Figure 4.2   An Mth-order FIR filter

The discrete-time counterpart to Dirac's delta function is called Kronecker's delta function $\delta(n)$. It is defined as follows

$$\delta(n) = \begin{cases} 1, & \text{if } n = 0 \\ 0, & \text{if } n \neq 0. \end{cases}$$

If the input to a discrete filter is Kronecker's delta function, then we obtain as the output the discrete-time pulse response $h(n)$. It is known that the output $y(n)$ of the discrete-time filter can be expressed via the input $x(n)$ and the discrete-time pulse response $h(n)$ as follows

$$y(n) = \sum_{m=0}^{n} x(m)h(n-m) = \sum_{m=0}^{n} h(m)x(n-m). \tag{4.2}$$

The sum in (4.2) is called a discrete-time *convolution*. A discrete-time filter with $h(n)$ of finite length, that is, $h(n) = 0$ for $n < n_0$ and $n > n_1$, where $n_0 < n_1$ is called a *Finite (Im)pulse Response* (FIR) filter. If in (4.1) we set $b_j = 0$, $j = 1, 2, \ldots, N$, then we obtain an equation which describes a FIR filter of order M. Its pulse response has nonzero samples only for $n = 0, 1, \ldots, M$. This filter can be realized with adders, multipliers, and unit delays as shown in Fig. 4.2. Since FIR filters do not contain feedback, they are also called *nonrecursive* filters. Notice that the order of the filter is equal to the number of its delay elements.

It follows from (4.1) that the output of the FIR filter is

$$y(n) = a_0 x(n) + a_1 x(n-1) + \cdots + a_M x(n-M). \tag{4.3}$$

For FIR filters formula (4.2) has the form

$$y(n) = \sum_{m=0}^{M} x(m)h(n-m) = \sum_{m=0}^{M} h(m)x(n-m). \tag{4.4}$$

Comparing (4.3) and (4.4), we obtain that $h(i) = a_i$, $i = 0, 1, \ldots, M$, that is, the samples of the pulse response are equal to coefficients a_i.

A filter with $h(n)$ of infinite length is referred to as an *Infinite (Im)pulse Response (IIR) filter*. In general, the *recursive* filter described by (4.1) represents an IIR filter.

In order to compute the pulse response of the discrete IIR filter, it is necessary to solve the corresponding recurrent equation (4.1). Consider the following example.

Example 4.1   The first-order IIR filter is described by the equation

$$y(n) = ax(n) + by(n-1). \tag{4.5}$$

It is evident that

$$y(0) = ax(0) + by(-1)$$

$$y(1) = b^2 y(-1) + abx(0) + ax(1).$$

Continuing in this manner, we obtain

$$y(n) = b^{n+1} y(-1) + \sum_{i=0}^{n} b^i ax(n-i).$$

By setting $y(-1) = 0$ and $x(n) = \delta(n)$, we obtain that $h(n) = ab^n$. The pulse response of the first-order IIR filter is depicted in Fig. 4.3 for different values of coefficients a and b.

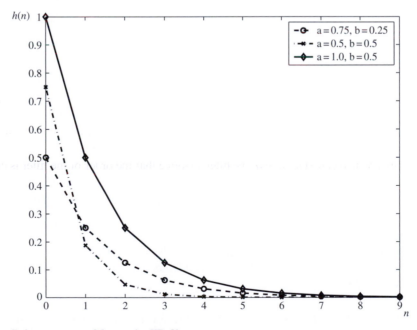

Figure 4.3   Pulse response of first-order IIR filter

4.1.2   Description by means of the z-transform

In order to analyze discrete-time filters, the so-called z-*transform* can be applied. The z-transform of a given discrete-time signal $x(n)$ is defined as

$$X(z) = \sum_{n=-\infty}^{\infty} x(n)z^{-n}, z \in E_x \tag{4.6}$$

where z is a complex variable, $X(z)$ denotes a function of this complex variable, and E_x is called the existence region. The existence region consists of all the complex numbers z for which the sum converges.

Example 4.2   Let us determine the z-transform of the causal sequence $x(n) = Ae^{-an}$ and its region of convergence. Applying the definition given by (4.6), we obtain

$$X(z) = A \sum_{n=-\infty}^{\infty} e^{-an}z^{-n} = A\sum_{n=0}^{\infty}(e^a z)^{-n}. \tag{4.7}$$

The second equality in (4.7) follows from the causality of $x(n)$. The power series (4.7) converges to

$$X(z) = \frac{A}{1 - e^{-a}z^{-1}} \text{ for } |e^{-a}z^{-1}| < 1.$$

Thereby, the existence region is $|z| > e^{-a}$.

The above definition is formulated for a noncausal signal, i.e. $x(n)$ is assumed to be known for time index $n < 0$. In most cases the signal will be causal and then the one-sided z-transform of sequence $x(0), x(1), \ldots, x(n)$ is defined as follows

$$X(z) = \sum_{n=0}^{\infty} x(n)z^{-n}, z \in E_x. \tag{4.8}$$

The z-transform for discrete signals plays the same role as the Laplace transform for analog signals. It allows us to replace recurrent equations over discrete sequences by algebraic equations over the corresponding z-transforms of these sequences.

Consider some properties of the z-transform.

Property 4.1   *(Linearity.) Let $X(z)$ and $Y(z)$ be the z-transforms of the sequences $x(n)$ and $y(n)$, respectively. Then the linear combination $ax(n) + by(n)$ has the z-transform $aX(z) + bY(z)$.*

Property 4.2   *(Discrete convolution.) Let $X(z)$ and $Y(z)$ be the z-transforms of the sequences $x(n)$ and $y(n)$, respectively. Then the product $R(z) = X(z)Y(z)$ is the z-transform of $r(n)$, where*

$$r(n) = \sum_{m=0}^{n} x(m)y(n-m) = \sum_{m=0}^{n} x(n-m)y(m)$$

that is, $r(n)$ represents a linear discrete convolution of the sequences $x(n)$ and $y(n)$.

Proof. Applying the definition of the z-transform to each of the co-factors of the product $X(z)Y(z)$, we obtain

$$\begin{aligned}
X(z)Y(z) &= \left(x(0) + x(1)z^{-1} + x(2)z^{-2} + \cdots\right) \\
&\quad \times \left(y(0) + y(1)z^{-1} + y(2)z^{-2} + \cdots\right) \\
&= x(0)y(0) + (x(0)y(1) + x(1)y(0))\, z^{-1} \\
&\quad + (x(0)y(2) + x(1)y(1) + x(2)y(0))\, z^{-2} + \cdots \\
&= r(0) + r(1)z^{-1} + r(2)z^{-2} + \cdots
\end{aligned}$$

By equating coefficients of like powers of z, we obtain that $r(n)$ represents the linear convolution of the sequences $x(n)$ and $y(n)$. $\square$

Property 4.3 *(Delay.) The z-transform of the delayed sequence $y(n) = x(n-m)$ is equal to $Y(z) = X(z)z^{-m}$, where $X(z)$ is the z-transform of $x(n)$.*

Proof. According to the definition

$$Y(z) = \sum_{n=0}^{\infty} y(n)z^{-n} = \sum_{n=0}^{\infty} x(n-m)z^{-n}. \tag{4.9}$$

Substituting $l = n - m$ into (4.9), we obtain

$$Y(z) = \sum_{l=-m}^{\infty} x(l)z^{-l}z^{-m} = \left(X(z) + \sum_{l=-m}^{-1} x(l)z^{-l}\right) z^{-m}.$$

If $x(n)$ is a causal sequence, then $\sum_{l=-m}^{-1} x(l)z^{-l} = 0$ and we obtain that the z-transform of $x(n-m)$ is equal to $X(z)z^{-m}$. $\square$

Examples of z-transforms for common sequences are presented in Table 4.1.

In the theory of linear digital filters we often deal with z-transforms which are rational functions of z^{-1}, that is, ratios of two polynomials in z^{-1}

$$X(z) = \frac{P(z)}{Q(z)} = \frac{p_0 + p_1 z^{-1} + p_2 z^{-2} + \cdots + p_M z^{-M}}{q_0 + q_1 z^{-1} + q_2 z^{-2} + \cdots + q_N z^{-N}} \tag{4.10}$$

$$= z^{N-M} \frac{p_0 z^M + p_1 z^{M-1} + \cdots + p_M}{q_0 z^N + q_1 z^{N-1} + \cdots + q_N}.$$

Equation (4.10) can be factorized as

$$X(z) = z^{N-M} \frac{p_0 \prod_{i=1}^{M}(z - \alpha_i)}{q_0 \prod_{i=1}^{N}(z - \beta_i)}$$

where α_i are called the *zeros* of $X(z)$ and β_i are called the *poles* of $X(z)$. Notice that $X(z)$ also has a multiple zero (of multiplicity $N - M$) at $z = 0$.

Table 4.1 z-transforms

Name of the sequence	Sequence	z-transform $X(z)$
Unit pulse	$x(n) = \begin{cases} 1, n = 0 \\ 0, n \neq 0 \end{cases}$	1
Unit step	$x(n) = \begin{cases} 1, n \geq 0 \\ 0, n < 0 \end{cases}$	$\frac{1}{1-z^{-1}}$
Exponent	$x(n) = Ae^{-\alpha n}$	$\frac{A}{1-e^{-\alpha}z^{-1}}$
Damped sine[a]	$x(n) = e^{-\alpha n} \sin(\omega n)$	$\frac{z^{-1}e^{-\alpha} \sin(\omega)}{B(z)}$
Damped cosine[a]	$x(n) = e^{-\alpha n} \cos(\omega n)$	$\frac{1-e^{-\alpha}z^{-1} \cos(\omega)}{B(z)}$
The second-order filter response	$x(n) = (\gamma_1^{n+1} - \gamma_2^{n+1})$, where $\gamma_{1,2} = a/2 \pm \sqrt{\frac{a^2}{4}+b}$ are real numbers	$\frac{1}{1-z^{-1}a-z^{-2}b} = \frac{1}{(z-\gamma_1)(z-\gamma_2)}$
Complex exponent	$x(n) = \begin{cases} e^{j\omega n}, n \geq 0 \\ 0, n < 0 \end{cases}$	$\frac{1}{1-e^{j\omega}z^{-1}}$

[a] $B(z) = 1 - z^{-1}2e^{-\alpha} \cos(\omega) + e^{-2\alpha}z^{-2}$

4.1.3 Inverse z-transform

There is a one-to-one correspondence between a discrete sequence and its z-transform. Let us consider how to compute the discrete sequence $x(n)$ when its z-transform $X(z)$ is known. According to the definition of the z-transform, $x(n)$ represents the inverse z-transform of $X(z)$. It can be found from (4.8) by using Cauchy's integral theorem (Churchill and Brown 1990). We multiply both sides of (4.8) by z^{k-1} and then take the contour integral of both sides of the equation. The contour integral is taken along a counterclockwise arbitrary closed path that encloses all the finite poles of $X(z)z^{k-1}$ and lies entirely in the existence region E_x. Since the path of integration lies entirely inside the domain of convergence for the infinite series (4.8), then the integration and summation can be interchanged which yields

$$\oint X(z)z^{k-1}dz = \sum_{n=0}^{\infty} x(n) \oint z^{k-n-1}dz. \tag{4.11}$$

It follows from Cauchy's theorem that if the path of integration encloses the origin ($z = 0$), then

$$\oint z^{k-n-1}dz = 0 \text{ for all } k \text{ except } k = n.$$

For $k = n$ it is equal to

$$\oint z^{-1} dz = 2\pi j.$$

Thus, expression (4.11) reduces to the following formula, which describes the inverse z-transform

$$x(k) = \frac{1}{2\pi j} \oint X(z) z^{k-1} dz. \tag{4.12}$$

Example 4.3 Let

$$X(z) = \frac{1}{\frac{1}{2} z^{-2} - \frac{3}{2} z^{-1} + 1}.$$

Decompose $X(z)$ into a sum of simple fractions

$$X(z) = \frac{1}{(1 - z^{-1}) \left(1 - \frac{1}{2} z^{-1}\right)} = \frac{A}{1 - z^{-1}} + \frac{B}{1 - \frac{1}{2} z^{-1}}. \tag{4.13}$$

Multiplying both sides of (4.13) by $(1 - z^{-1}) \left(1 - \frac{1}{2} z^{-1}\right)$ we obtain

$$1 = A \left(1 - \frac{1}{2} z^{-1}\right) + B(1 - z^{-1}). \tag{4.14}$$

By equating coefficients of like powers of z, we obtain

$$\begin{cases} A + B = 1 \\ -\frac{A}{2} - B = 0. \end{cases} \tag{4.15}$$

Solving the system of linear equations (4.15), we obtain

$$X(z) = \frac{1}{(1 - z^{-1}) \left(1 - \frac{1}{2} z^{-1}\right)} = \frac{2}{1 - z^{-1}} - \frac{1}{1 - \frac{1}{2} z^{-1}}. \tag{4.16}$$

Comparing (4.16) with the z-transform pairs given in Table 4.1 we find that the first summand on the right side of (4.16) corresponds to

$$x(n) = \begin{cases} 2, & \text{if } n \geq 0 \\ 0, & \text{if } n < 0 \end{cases}$$

and the second summand corresponds to $x(n) = -2^{-n}$. Thus, we obtain

$$x(n) = \begin{cases} 2 - 2^{-n}, & \text{if } n \geq 0 \\ 0, & \text{if } n < 0. \end{cases}$$

In the case of rational z-transforms the contour integral in (4.12) can be evaluated via the theorem of residues (Churchill and Brown 1990):

Theorem 4.1   *Let function $f(z)$ be unique and analytic (differentiable) every-where along the contour L and inside it except in the so-called singular points z_k, $k = 1, \ldots, N$, lying inside L. Then*

$$\oint_{(L)} f(z)dz = 2\pi j \sum_{k=1}^{N} Res[f(z), z_k]$$

that is, the contour integral is equal to the sum of residues $Res\left[f(z), z_k\right]$, $k = 1, 2, \ldots, N$ of integrand function taken in the singular points lying inside the path of integration, multiplied by $2\pi j$.

Assume that $f(z)$ has the pole $z = z_0$ of multiplicity m; that is, it can be represented in the form

$$f(z) = \frac{P(z)}{(z - z_0)^m}$$

where $P(z)$ does not have a zero in z_0. Then, the residue in the mth-order pole z_0 can be evaluated as

$$Res[f(z), z_0] = \frac{1}{(m-1)!} \lim_{z \to z_0} \frac{d^{m-1}}{dz^{m-1}} \left[(z - z_0)^m f(z)\right].$$

If the function $f(z)$ represents a ratio of two finite functions $p(z)$ and $q(z)$, that is, $f(z) = p(z)/q(z)$ and it is known that $q(z)$ has zero of the first order in $z = a$, and that $p(a) \neq 0$, then the residue in $z = a$ can be found as

$$Res[f(z), a] = \frac{p(a)}{dq(z)/dz|_{z=a}}. \tag{4.17}$$

Thus, if the function $X(z)z^{n-1}$ has K finite poles at $z = a_i$, $i = 1, 2, \ldots, K$, then the integration formula reduces to

$$x(n) = \sum_{i=1}^{K} Res\left[X(z)z^{n-1}, z = a_i\right].$$

Example 4.4   Continuing Example 4.3 we obtain that

$$x(n) = \oint \frac{2z^{n-1}dz}{1 - z^{-1}} - \oint \frac{z^{n-1}dz}{1 - \frac{1}{2}z^{-1}}$$

and using formula (4.17) that

$$Res\left[\frac{2z^n}{z-1}, 1\right] = \frac{2}{1}, \ Res\left[\frac{2z^n}{2z-1}, \frac{1}{2}\right] = \frac{2(1/2)^n}{2} = 2^{-n}.$$

Thus, $x(n) = 2 - 2^{-n}$.

Let us solve equation (4.5) using z-transform. By using properties of z-transform (linearity and delay), we obtain

$$Y(z) = bY(z)z^{-1} + aX(z)$$

$$H(z) = \frac{Y(z)}{X(z)} = \frac{a}{1 - bz^{-1}}$$

where $H(z)$ is the transfer function of the discrete-time filter, and $X(z)$ and $Y(z)$ are the z-transforms of the input and output sequences, respectively. The transfer function $H(z)$ represents the z-transform of the pulse response $h(n)$; that is, it can be written as

$$H(z) = \sum_{n=0}^{\infty} h(n)z^{-n}.$$

It is evident that $Y(z) = X(z)H(z)$ and, hence, if we know the z-transform of the input sequence we can evaluate the z-transform of the output sequence. If $x(n) = \delta(n)$, then $X(z) = 1$. This yields

$$Y(z) = H(z) = \frac{a}{1 - bz^{-1}} = \frac{az}{z - b}.$$

By applying the inverse z-transform to $H(z)$, we obtain $y(n) = ab^n$ if the integration path encloses the pole $z = b$. This coincides with the result obtained by solving equation (4.5).

Notice that the stability condition for the discrete-time linear filter can be reformulated via its transfer function. We say that the filter with the transfer function $H(z)$ is stable if the poles of $H(z)$ lie inside the unit circle in the z-plane.

4.1.4 Frequency function

The complex frequency function of the discrete-time linear filter can be obtained by inserting $z = e^{j\omega T_s}$ into the transfer function $H(z)$. Consider the discrete-time signal $x(nT_s) = e^{j\omega nT_s}$. The corresponding output from the discrete-time linear system with pulse response $h(nT_s)$ is

$$y(nT_s) = \sum_{m=0}^{n} h(mT_s)x(nT_s - mT_s)$$

$$= \sum_{m=0}^{n} h(mT_s)e^{j\omega(n-m)T_s}$$

$$= e^{j\omega nT_s} \sum_{m=0}^{n} h(mT_s)e^{-j\omega mT_s}.$$

When $n \to \infty$, we obtain that

$$y(nT_s) = e^{j\omega nT_s} \sum_{m=0}^{\infty} h(mT_s)e^{-j\omega mT_s} = e^{j\omega nT_s} H\left(e^{j\omega T_s}\right)$$

where

$$H\left(e^{j\omega T_{\mathrm{s}}}\right) = \sum_{m=0}^{\infty} h(mT_{\mathrm{s}})e^{-j\omega mT_{\mathrm{s}}}$$

is the complex *frequency function* of the discrete-time linear filter.

The complex frequency function can be represented in the form

$$H(e^{j\omega T_{\mathrm{s}}}) = A(\omega)e^{j\varphi(\omega)} = Re\left\{H(e^{j\omega T_{\mathrm{s}}})\right\} + jIm\left\{H(e^{j\omega T_{\mathrm{s}}})\right\}$$

where

$$A(\omega) = \sqrt{\{Re(H(e^{j\omega T_{\mathrm{s}}}))\}^2 + \{Im(H(e^{j\omega T_{\mathrm{s}}}))\}^2}$$

is the *amplitude function* and

$$\varphi(\omega) = \arctan\frac{Im(H(e^{j\omega T_{\mathrm{s}}}))}{Re(H(e^{j\omega T_{\mathrm{s}}}))}$$

is the *phase function*. If the filter output is the discrete-time sinusoid $x(nT_{\mathrm{s}}) = \sin(n\omega T_{\mathrm{s}})$, then the amplitude function and the phase function describe the change in amplitude and phase, respectively, which is introduced by the discrete-time linear filter. Thus, the output sine signal looks like

$$y(nT_{\mathrm{s}}) = A(\omega)\sin(n\omega T_{\mathrm{s}} + \varphi(\omega)).$$

The frequency function has the following properties:

- The frequency function is a continuous function of the frequency.
- The frequency function is a periodic function of the frequency with the period equal to the sampling frequency.
- For discrete-time linear filters with coefficients a_i and b_i, which are real numbers, the amplitude function is an even function of frequency and the phase function is an odd function of frequency.
- The frequency function can be represented via the pulse response as

$$H(e^{j\omega T_{\mathrm{s}}}) = \sum_{n=0}^{\infty} h(nT_{\mathrm{s}})e^{-j\omega nT_{\mathrm{s}}}$$

$$= \sum_{n=0}^{\infty} h(nT_{\mathrm{s}})e^{-j2\pi nf T_{\mathrm{s}}}$$

$$= \sum_{n=0}^{\infty} h(nT_{\mathrm{s}})(\cos(2\pi nf T_{\mathrm{s}}) - j\sin(2\pi nf T_{\mathrm{s}})).$$

When comparing different frequency functions it is often convenient to consider the so-called normalized frequency $\alpha = \omega/\omega_{\mathrm{s}}$, where $\omega_{\mathrm{s}} = 2\pi/T_{\mathrm{s}}$ is the sampling frequency in radians. Then, we obtain

$$H(e^{j2\pi\alpha}) = \sum_{n=0}^{\infty} h(nT_{\mathrm{s}})e^{-j2\pi n\alpha} = \sum_{n=0}^{\infty} h(nT_{\mathrm{s}})(\cos(2\pi n\alpha) - j\sin(2\pi n\alpha)).$$

Example 4.5 Consider a discrete-time linear filter of the first order. It is described by equation (4.5). Its transfer function has the form

$$H(z) = \frac{a}{1 - bz^{-1}}.$$

By inserting $z = e^{j\omega T_s}$ into the expression for the transfer function, we obtain its frequency function. It has the form

$$H(e^{j\omega T_s}) = \frac{a}{1 - be^{-j\omega T_s}} = \frac{a}{1 - b\cos(\omega T_s) + jb\sin(\omega T_s)}.$$

The amplitude function is

$$A(\omega) = \frac{a}{\sqrt{(1 - b\cos(\omega T_s))^2 + b^2 \sin^2(\omega T_s)}} = \frac{a}{\sqrt{1 - 2b\cos(\omega T_s) + b^2}}$$

and the phase function is

$$\varphi(\omega) = -\arctan\frac{b\sin(\omega T_s)}{1 - b\cos(\omega T_s)}.$$

By normalizing the frequency we obtain the following formulas for the amplitude and phase functions

$$A(\alpha) = \frac{a}{\sqrt{1 - 2b\cos(2\pi\alpha) + b^2}}$$

$$\varphi(\alpha) = -\arctan\frac{b\sin(2\pi\alpha)}{1 - b\cos(2\pi\alpha)}.$$

The amplitude function $A(\alpha)$ for $a = 1$, $b = 0.5$ is shown in Fig. 4.4.

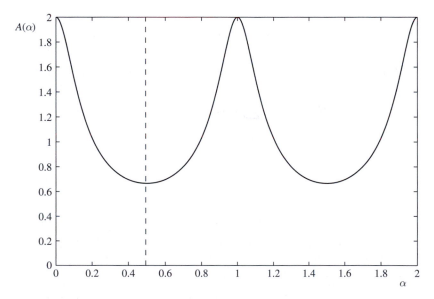

Figure 4.4 The amplitude function of the first-order discrete-time linear filter with $a = 1$ and $b = 0.5$

From the properties of the frequency function it follows that only the first half-period of $A(\alpha)$ $(0 \leq \alpha \leq 1/2)$ contains information about the filter performances. It is easy to see in Fig. 4.4 that the first half-period of $A(\alpha)$ characterizes a lowpass filter.

4.2   Linear predictive coding

As was mentioned above, linear predictive coding combines *linear processing with scalar quantization*. It removes source redundancy by using a linear prediction method. The main idea of this method is to predict the value of the current sample by a linear combination of the previous (true or already reconstructed) samples and then to quantize the difference between the actual value and the predicted value. Linear prediction coefficients are weighting coefficients used in a linear combination. A simple predictive quantizer or *differential pulse-coded modulator* is depicted in Fig. 4.5. It compares the input sample $x(n)$ with its prediction $\hat{x}(n)$ and quantizes their difference $e(n)$. If the prediction is simply the last sample (or, more precisely, its approximation) and the quantizer generates only one bit per sample, the system becomes a *delta-modulator*. It is illustrated in Fig. 4.6, where approximation of the last sample is generated by

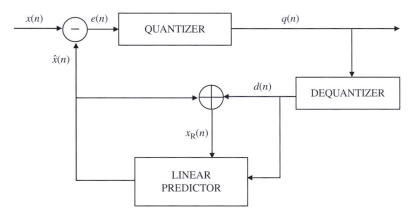

Figure 4.5   Differential pulse-coded modulator

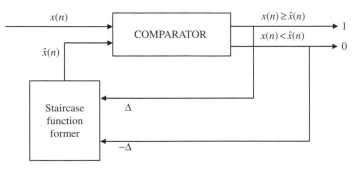

Figure 4.6   Delta-modulator

a staircase function former. Since samples of such signals as speech and images are highly correlated, then the prediction error is usually rather small; that is, the primary role of the predictor is to reduce the variance of the variable to be scalar-quantized. Notice that the main feature of the quantizers shown in Figs 4.5 and 4.6 is that they exploit not all the advantages of predictive coding. In particular, the linear prediction coefficients used in these schemes are not optimal and the prediction is based on the past reconstructed samples and not the true samples. Usually, the coefficients of prediction are chosen by using some empirical rules and are not transmitted. For example, the quantizer shown in Fig. 4.5 uses approximating values $d(n)$ reconstructed in the dequantizer instead of the actual value of prediction error $e(n)$ and estimates $x_R(n)$ obtained via $d(n)$ which are not exactly equal to $e(n)$ instead of the true sample values $x(n)$. As a result, the compression systems based on differential pulse-code modulating provide high quality synthesized speech for rather high bit-rates and correspondingly small compression ratios. The main advantage of these systems is their low complexity.

The most advanced quantizer of the linear predictive type represents a basis of the so-called Code Excited Linear Predictive (CELP) coder and it uses the optimal set of coefficients or, in other words, the linear prediction coefficients of this quantizer are determined by minimizing the mean squared prediction error between the current speech fragment and its prediction. Moreover, the prediction is based on the original past samples. Notice that using the true samples for the prediction requires the so-called "looking ahead" procedure in the coder and the prediction coefficients have to be known at the decoder side.

Predictive quantizers have been extensively developed. There are many adaptive versions, and they are widely used in speech coding, where a number of standards are based on them. For example, they form the basis of the ITU standards G721, G722, G723, and G726 standards which will be discussed in Chapter 7.

Now we will consider the linear predictive quantizer in more detail. Let $x(1)$, $x(2)$, ... be a sequence of samples at the quantizer input. We assume that quantizer coefficients are optimized for a group of samples and that the original past samples are used for prediction. Then, each sample $x(n)$ is predicted by the previous samples according to the formula

$$\hat{x}(n) = \sum_{k=1}^{m} a_k x(n-k) \tag{4.18}$$

where $\hat{x}(n)$ denotes the predicted value of the nth sample, a_k are the prediction coefficients, and m is the order of prediction. The prediction error is determined as follows

$$e(n) = x(n) - \hat{x}(n). \tag{4.19}$$

Since speech is a nonstationary signal, we shall find the prediction coefficients which minimize the sum of squared prediction errors over a given finite interval considered as the *interval of quasi-stationarity*. Such an interval is called the *frame* of the speech signal.

Let $[n_0, n_1]$ be such an interval; then the sum of the squared prediction errors is calculated as follows:

$$E = \sum_{n=n_0}^{n_1} e^2(n). \tag{4.20}$$

By inserting (4.18) into (4.20), we obtain

$$E = \sum_{n=n_0}^{n_1} \left(x(n) - \sum_{k=1}^{m} a_k x(n-k) \right)^2$$

$$= \sum_{n=n_0}^{n_1} \{x(n)\}^2 - 2 \sum_{j=1}^{m} a_j \sum_{n=n_0}^{n_1} x(n)x(n-j)$$

$$+ \sum_{j=1}^{m} \sum_{k=1}^{m} a_j a_k \sum_{n=n_0}^{n_1} x(n-j)x(n-k). \tag{4.21}$$

Differentiating (4.21) over a_k, $k = 1, 2, \ldots, m$, yields

$$\partial E / \partial a_k = \sum_{n=n_0}^{n_1} x(n)x(n-k) - \sum_{j=1}^{m} a_j \sum_{n=n_0}^{n_1} x(n-k)x(n-j) = 0.$$

Thus, we obtain a system of m linear equations with m unknown quantities $a_1, a_2, \ldots, a_m$:

$$\sum_{j=1}^{m} a_j c_{jk} = c_{0k}, \quad k = 1, 2, \ldots, m \tag{4.22}$$

where

$$c_{jk} = c_{kj} = \sum_{n=n_0}^{n_1} x(n-j)x(n-k). \tag{4.23}$$

This system of linear equations is called the *Yule–Walker prediction equations*. If a_1, a_2, ..., a_m are the solutions of (4.22), then we can evaluate the minimal achievable prediction error $E_{\min}$. By inserting (4.23) into (4.21), we obtain

$$E_{\min} = c_{00} - 2 \sum_{k=1}^{m} a_k c_{0k} + \sum_{k=1}^{m} a_k \sum_{j=1}^{m} a_j c_{jk}. \tag{4.24}$$

Using (4.22) we reduce (4.24) to the expression

$$E_{\min} = c_{00} - \sum_{k=1}^{m} a_k c_{0k}. \tag{4.25}$$

It is easy to see that equation (4.18) describes the mth-order predictor whose transfer function is equal to

$$P(z) = \frac{\hat{X}(z)}{X(z)} = \sum_{k=1}^{m} a_k z^{-k}. \tag{4.26}$$

It follows from (4.18), (4.19), and (4.26) that the z-transform for the prediction error has the form

$$E(z) = X(z) - \sum_{k=1}^{m} a_k X(z) z^{-k}.$$

In other words, the prediction error is an output signal of the discrete-time system with transfer function

$$H_e(z) = \frac{E(z)}{X(z)} = 1 - \sum_{k=1}^{m} a_k z^{-k}.$$

It is said that the problem of finding the optimal set of linear prediction coefficients is reduced to the problem of constructing the optimal prediction filter of order m. This filter represents a discrete-time FIR filter.

Another name of the linear prediction (4.18) is the *autoregressive model* of the discrete-time signal $x(n)$. The signal $x(n)$ can be generated according to the recurrent equation

$$x(n) = e(n) + \sum_{k=1}^{m} a_k x(n-k),$$

that is, as the output of the so-called *autoregressive filter* with transfer function

$$H_a(z) = \frac{1}{\left(1 - \sum\limits_{k=1}^{m} a_k z^{-k}\right)}.$$

This filter is the inverse with respect to the prediction filter. It is evident that it is a discrete-time IIR filter.

In order to find the optimal set of prediction coefficients $a_1, a_2, \ldots, a_m$, it is necessary to solve the Yule–Walker equations (4.22). The coefficients c_{ij}, $i = 0, 1, \ldots, m$, $j = 0, 1, \ldots, m$ of (4.22) are estimated for each frame of the input signal. There are two procedures producing these estimates. Moreover, it will be shown that the computational complexity of solving equations (4.22) depends on how the coefficients c_{ij} are determined. The first method is called the autocorrelation method and the second one is called the covariance method (Rabiner and Schafer 1978).

4.2.1   Autocorrelation method

The values c_{ij} are computed as

$$c_{ij} = c_{ji} = \sum_{n=n_0}^{n_1} x(n-i)x(n-j).$$

Let N be the frame length. To simplify the notations we assign the zero time index to the first sample of the frame. The autocorrelation method implies that $x(n) = 0$, if $N-1 < n < 0$. In other words, the samples outside of the frame are set to 0. In this case we can simplify the expression for c_{ij}, $i = 1, 2, \ldots, m$, $j = 0, 1, \ldots, m$, to

$$c_{ij} = \sum_{n=0}^{N-1-|i-j|} x(n)x(n+|i-j|). \tag{4.27}$$

Figure 4.7 demonstrates the frame of the speech signal $x(n)$ of size $N = 60$ and its shift by 10 samples for the autocorrelation method. It is easy to see that the product $x(n)x(n+|i-j|)$ is nonzero only for the interval from $n = 20$ up to $n = 69$; that is, for $N - |i-j| - |i-j|$ samples. Assuming that $x(n)$ is a discrete-time stationary signal of length N with zero mean, we conclude that the value c_{ij} normalized by N coincides with an estimate of the entry $R(|i-j|)$ of the autocorrelation matrix

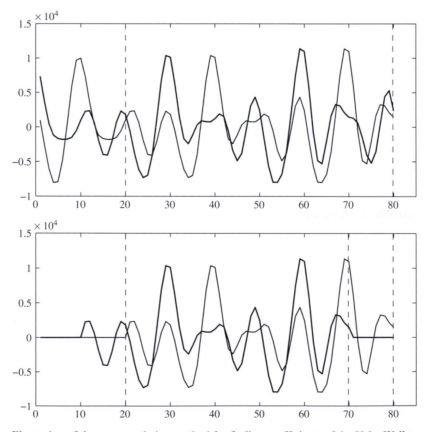

Figure 4.7   Illustration of the autocorrelation method for finding coefficients of the Yule–Walker equations

$$\hat{R}(|i - j|) = c_{ij}/N = 1/N \sum_{n=0}^{N-1-|i-j|} x(n)x(n + |i - j|).$$

Notice that for discrete-time signals with zero mean, the autocorrelation matrix coincides with the covariance matrix. In general, the covariance values are computed for centered random variables whereas the correlation values are computed for noncentered random variables.

Dividing equations (4.22) by N, we obtain the Yule–Walker equations for the autocorrelation method

$$\sum_{i=1}^{m} a_i \hat{R}(|i - j|) = \hat{R}(j) , \, j = 1, 2, \ldots, m. \tag{4.28}$$

The system (4.28) can be given by the matrix equation

$$a \times \mathbf{R} = b$$

where $a = (a_1, a_2, \ldots, a_m)$, $b = (\hat{R}(1), \hat{R}(2), \ldots, \hat{R}(m))$,

$$\mathbf{R} = \begin{pmatrix} \hat{R}(0) & \hat{R}(1) & \ldots & \hat{R}(m - 1) \\ \hat{R}(1) & \hat{R}(0) & \ldots & \hat{R}(m - 2) \\ \ldots & \ldots & \ldots & \ldots \\ \hat{R}(m - 1) & \hat{R}(m - 2) & \ldots & \hat{R}(0) \end{pmatrix}.$$

The matrix $\mathbf{R}$ of the autocorrelation method has two important properties. It is symmetric, that is, $\hat{R}(i, j) = \hat{R}(j, i)$, and it has the Toeplitz property, that is,

$$\hat{R}(i, j) = \hat{R}(|i - j|).$$

The Toeplitz property of the matrix $\mathbf{R}$ makes it possible to simplify the solution of (4.22). For example, the Levinson–Durbin fast recursive algorithm, which is considered below, requires only m^2 operations. Notice that the computational complexity of solving an arbitrary system of linear equations of order m is equal to m^3 operations.

4.2.2 Covariance method

We choose $n_0 = 0$ and $n_1 = N - 1$, and assume that the signal $x(n)$ is not constrained in time. In this case the values c_{ij}, $i = 0, 1, \ldots, m$, $j = 1, 2, \ldots, m$, can be expressed as follows

$$c_{ij} = \sum_{n=0}^{N-1} x(n - i)x(n - j). \tag{4.29}$$

Set $k = n - i$; then we can represent (4.29) in the form

$$c_{ij} = \sum_{k=-i}^{N-i-1} x(k)x(k + i - j) \tag{4.30}$$

$i = 1, 2, \ldots, m$, $j = 0, 1, \ldots, m$.

Expression (4.30) resembles expression (4.27) used by the autocorrelation method, but it has another range of definition for index k. It is evident that (4.30) uses signal values $x(k)$ outside of the range $0 \le k \le N - 1$. In other words, in order to use the covariance method for computing c_{ij}, we have to know the signal values $x(-m), x(-m + 1), \ldots, x(N - 1)$; that is, we need to know the signal for interval $N + m$ samples instead of N samples as for the autocorrelation method. Usually, this is not a critical point since $m << N$. This method leads to the so-called cross-correlation matrix for two similar but not identical finite fragments of signal $x(k)$. For example, if $i = m$ and $j = 0$, we obtain that c_{m0} represents the correlation value between the fragment $x(-m), x(-m + 1), \ldots, x(N - 1 - m)$ and the fragment $x(0), x(1), \ldots,$ $x(N - 1)$, which differ from each other in the samples $x(-m), x(-m + 1), \ldots, x(-1)$ and cannot be considered as a signal and its shift by m samples. In this case the values c_{ij} normalized by N coincide with an estimate of the entry $\hat{R}(i, j)$ of the cross-correlation matrix

$$\hat{R}(i, j) = c_{ij}/N = 1/N \sum_{n=0}^{N-1} x(n - i)x(n - j).$$

It is easy to see that $\hat{R}(i, j) = \hat{R}(j, i)$ but $\hat{R}(i, j)$ is not a function of $i - j$ as it is for the autocorrelation method. Dividing the equations (4.22) by N, we obtain the Yule–Walker equations for the covariance method

$$\sum_{i=1}^{m} a_i \hat{R}(i, j) = \hat{R}(0, j) , j = 1, 2, \ldots, m. \tag{4.31}$$

The equations (4.31) can be written as a matrix equation:

$$\boldsymbol{a} \times \mathbf{P} = \boldsymbol{c}$$

where $\boldsymbol{a} = (a_1, a_2, \ldots, a_m)$, $\boldsymbol{c} = (\hat{R}(0, 1), \hat{R}(0, 2), \ldots, \hat{R}(0, m))$, and

$$\mathbf{P} = \begin{pmatrix} \hat{R}(1, 1) & \hat{R}(1, 2) & \ldots & \hat{R}(1, m) \\ \hat{R}(2, 1) & \hat{R}(2, 2) & \ldots & \hat{R}(2, m) \\ \ldots & \ldots & \ldots & \ldots \\ \hat{R}(m, 1) & \hat{R}(m, 2) & \ldots & \hat{R}(m, m) \end{pmatrix}.$$

The matrix $\mathbf{P}$ is symmetric, but unlike the matrix $\mathbf{R}$ of the autocorrelation method, $\mathbf{P}$ is a nonToeplitz matrix. In general, m^3 operations are required to solve (4.31). Notice that this procedure uses the last samples of the previous frame in order to predict the first samples of the current frame and does not require the artificial zero-setting. It might provide better performances than the autocorrelation method. As mentioned in (Jayant and Noll 1984), the terms *autocorrelation* and *covariance* are not apt, since the difference between the two procedures is not related to the difference between these two terms.

4.2.3   **Algorithms for the solution of the Yule–Walker equations**

The computational complexity of solving the Yule–Walker equations depends on the method of evaluating c_{ij}. Assume that c_{ij} are found by the autocorrelation method. In this case the Yule–Walker equations have the form (4.28) and the matrix $\mathbf{R}$ is both symmetric and Toeplitz. Due to these properties, (4.28) may be solved by fast recursive algorithms requiring only m^2 operations. There are a few methods of this type: the Levinson–Durbin algorithm, the Euclidean algorithm, the Berlekamp–Massey algorithm.

Consider the so-called Levinson–Durbin recursive algorithm. It was suggested by Levinson in 1947 (Levinson 1947) and improved by Durbin in 1960 (Durbin 1960). Notice that this algorithm works efficiently if the matrix $\mathbf{R}$ is simultaneously symmetric and Toeplitz. As for the Berlekamp–Massey and Euclidean algorithms, they do not require the matrix $\mathbf{R}$ to be symmetric.

We solve the equations (4.28) of order $l = 1, 2, \ldots, m$ sequentially. Let $\boldsymbol{a}^{(l)} = (a_1^{(l)}, a_2^{(l)}, \ldots, a_l^{(l)})$ denote the solution for the system of order l. Given $\boldsymbol{a}^{(l)}$ we find the solution for order $(l+1)$. At each step of the algorithm we obtain the prediction error E_l of the system of order l. Notice that rather often, instead of the prediction error, we estimate the *relative prediction error*, that is, the ratio $E_l / \sum_{i=1}^{N} x_i^2 / N = E_l / \hat{R}(0)$. At the last step of the algorithm, that is, when $l = m$, we obtain the solution

$$\boldsymbol{a} = (a_1, a_2, \ldots, a_m) = \boldsymbol{a}^{(m)}, \ E = E_m.$$

The formal description of the algorithm is given below.

Input:
order m
entries $R(0), \ldots, R(m)$ of the covariance matrix
Output:
filter coefficients $\boldsymbol{a} = (a_1, a_2, \ldots, a_m)$

Initialization: $l = 0$, $E_0 = R(0)$, $a^{(0)} = 0$

for $l = 1$ **to** m **do**

$$a_l^{(l)} = \frac{\sum_{i=1}^{l-1} \left(R(l) - a_i^{(l-1)} R(l-i) \right)}{E_{l-1}} \tag{4.32}$$

$$a_j^{(l)} = a_j^{(l-1)} - a_l^{(l)} a_{l-j}^{(l-1)}, \quad 1 \le j \le l-1$$

$$E_l = E_{l-1} \left(1 - \left\{ a_l^{(l)} \right\}^2 \right). \tag{4.33}$$

end
At the last step of the algorithm we obtain the solution

$$\boldsymbol{a} = (a_1, a_2, \ldots, a_m) = \boldsymbol{a}^{(m)}, \quad E = E_m.$$

Algorithm 4.1   Levinson–Durbin algorithm

It is easy to estimate the computational complexity of the Levinson–Durbin algorithm. It follows from (4.32) and (4.33) that the complexity of each step of the procedure is roughly proportional to the step number, that is, the total number of operations is proportional to

$$1 + 2 + \cdots + m = \frac{m(m+1)}{2}.$$

Example 4.6 Consider a speech frame of size $N = 240$ as shown in Fig. 4.8. The first 20 samples of this frame are: 2333, 682, -1915, -3985, -4075, -2074, 706, 2314, 1821, -550, -3694, -6134, -7306, -6971, -4083, 1229, 6827, 10369, 10130, 6118.

Assume that for the given speech frame, we are going to construct a prediction filter of order 2. By applying the autocorrelation method, we find the following estimates of the entries of the covariance matrix:

$$\hat{R}(0) = 3.79 \cdot 10^7, \ \hat{R}(1) = 2.85 \cdot 10^7, \ \hat{R}(2) = 6.62 \cdot 10^6.$$

The coefficients of the Yule–Walker equations normalized by $\hat{R}(0)$ are:

$$1.0, 0.7524, 0.1747.$$

Thus, the Yule–Walker equations have the form

$$\begin{cases} a_1 + a_2 0.7524 = 0.7524 \\ a_1 0.7524 + a_2 = 0.1747. \end{cases}$$

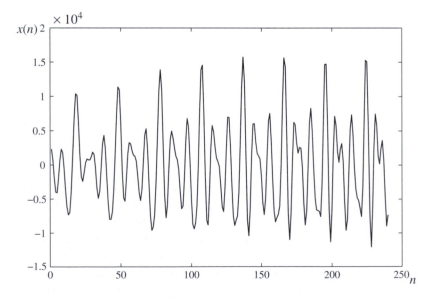

Figure 4.8   A speech frame of size $N = 240$

Consider two steps of the Levinson–Durbin procedure.
Step 1.

$$a_1^{(1)} = \hat{R}(1)/\hat{R}(0) = 0.7524$$

$$E_1/\hat{R}(0) = \left(1 - \left\{a_1^{(1)}\right\}^2\right) = 0.434.$$

Step 2.

$$a_2^{(2)} = \left(\hat{R}(2) - a_1^{(1)}\hat{R}(1)\right)/E_1 = -0.9021$$

$$a_1^{(2)} = a_1^{(1)}(1 - a_2^{(2)}) = 1.431$$

$$E_2/\hat{R}(0) = E_1/\hat{R}(0)\left(1 - \left\{a_2^{(2)}\right\}^2\right) = 0.074.$$

The prediction filter is described by the following recurrent equation

$$e(n) = x(n) - 1.431x(n-1) + 0.9021x(n-2).$$

Its transfer function has the form

$$H_e(z) = E(z)/X(z) = 1 - 1.431z^{-1} + 0.9021z^{-2}.$$

The amplitude function of the prediction filter is

$$A_e(\alpha) = \sqrt{H_{eR}^2(\alpha) + H_{eI}^2(\alpha)}$$

where

$$H_{eR}(\alpha) = 1 - 1.431\cos(2\pi\alpha) + 0.9021\cos(4\pi\alpha)$$

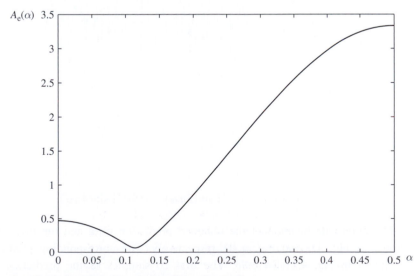

Figure 4.9   Amplitude function of prediction filter of order 2

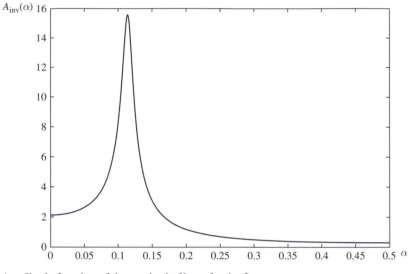

Figure 4.10   Amplitude function of the synthesis filter of order 2

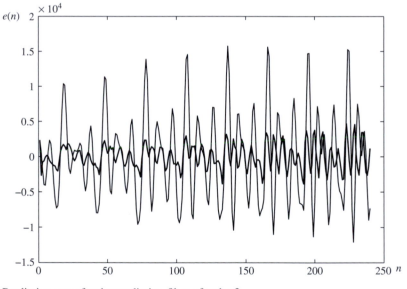

Figure 4.11   Prediction error for the prediction filter of order 2

$$H_{\mathrm{eI}}(\alpha) = 1.431 \sin(2\pi\alpha) - 0.9021 \sin(4\pi\alpha).$$

The amplitude function of the obtained prediction filter and the inverse filter are shown in Fig. 4.9 and Fig. 4.10, respectively. The corresponding prediction error is shown in Fig. 4.11, in bold. The first 20 samples of the prediction error are: 2333.0, −2660.0, −786.0, −620.0, −98.7, 163.0, −248.0, −568.0, −854.0, −1070.0,

$-1260.0, -1340.0, -1860.0, -2050.0, -697.0, 783.0, 1380.0, 1710.0, 1450.0, 974.0$.
It is easy to see that the dynamic range of the prediction error signal is significantly narrower than for the original speech frame.

Problems

4.1 Let the transfer function of the discrete-time filter be

$$H(z) = \frac{1}{\frac{1}{3}z^{-2} - \frac{4}{3}z^{-1} + 1}.$$

Find the pulse response of the filter.

4.2 Find the pulse response and the frequency function for the discrete-time linear filter of order 2 described by the recurrent equation

$$y(n) = b_1 y(n-1) + b_2 y(n-2) + x(n)$$

$$y(-1) = y(-2) = 0.$$

4.3 Find the amplitude and the phase of the discrete-time linear filter given in Problem 4.2. Plot the amplitude for $b_1 = 0.80$, $b_2 = 0.20$; and $b_1 = 0.80$, $b_2 = -0.52$.

4.4 Draw a block diagram of the discrete-time linear filter described by the recurrent equation

$$y(n) = b_1 y(n-1) + b_2 y(n-2) + b_3 y(n-3) + a_0 x(n) + a_1 x(n-1).$$

4.5 Describe the filter given in Problem 4.4 using the z-transform.

4.6 Write down the recurrent equation for the filter with the transfer function

$$H(z) = \frac{z^{-1} - a}{(z^{-1} - b)(z^{-1} - c)}.$$

4.7 Assume that the first-order predictor is used, i.e.

$$\hat{y}(n) = ay(n-1).$$

Derive the optimal value of the prediction coefficient a.

4.8 Let the coefficients of the Yule–Walker equations of order 4 be equal to $R(0) = 3.79 \times 10^7$, $R(1) = 2.85 \times 10^7$, $R(2) = 6.62 \times 10^6$, $R(3) = -1.46 \times 10^7$, and $R(4) = -2.48 \times 10^7$.

Write down the Yule–Walker equations of the given order. Find the solution of the Yule–Walker equations by using the Levinson–Durbin procedure. Estimate the mean squared prediction error.

4.9   The coefficients of the Yule–Walker equations of order 2 are: $R(0) = 4.39 \times 10^7$, $R(1) = 2.89 \times 10^7$, and $R(2) = -1.82 \times 10^6$.

Find the solution of the Yule–Walker equations by using the Levinson–Durbin procedure. Write down the equation for the prediction filter. Write down the transfer function of the prediction filter. Write down the equation for the synthesis filter. Write down the transfer function of the synthesis filter.

5 Transform coding

Transform coding (Jayant and Noll 1984) is the second approach to exploiting redundancy (source memory) by using scalar quantization with linear preprocessing. The source samples x_i are collected into a vector of dimension N that is linearly transformed into a vector of N transform coefficients $\boldsymbol{y} = (y_1, y_2, \ldots, y_N)$. Such a linear transform can be described as a multiplication of the input vector $\boldsymbol{x} = (x_1, x_2, \ldots, x_N)$ by a *transform matrix* T of size $N \times N$. The value N is referred to as the *order of the transform*. The obtained coefficients y_i, $i = 1, 2, \ldots, N$, are then quantized. Scalar quantization of transform coefficients is rather efficient but, in principle, they can be vector quantized with lower average distortions. Notice that if scalar quantization is used, then each coefficient can be quantized by a different quantizer. In the decoder the inverse transform described by the matrix T^{-1} is applied to the vector of approximating values $\hat{\boldsymbol{y}} = (\hat{y}_1, \hat{y}_2, \ldots, \hat{y}_N)$. A block diagram of transform coding is shown in Fig. 5.1. This kind of lossy coding was introduced in 1956 by Kramer and Mathews (1956), and analyzed and popularized in 1962–3 by Huang and Schultheiss (1963). This method has been developed for coding of images and video, where the Discrete Cosine Transform (DCT) is most commonly used because of its good performance. Indeed, DCT coding is the basic approach dominating current image and video coding standards, including H.261, H.263, JPEG, and MPEG. More recently, transform coding has also been widely used in high-fidelity audio coding.

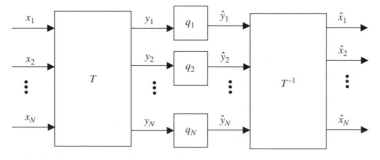

Figure 5.1   A block diagram of transform coding

5.1   Properties of transforms

We consider only invertible transforms. Thus, a transform itself does not introduce any distortions into x and if we remove the quantization block in Fig. 5.1, then $\hat{x} = x$. Usually it is required that the transform would have the following properties:

- localization of the essential part of the signal energy in a small number of transform coefficients. It makes it possible after quantization to exclude the least informative coefficients from consideration;
- transform coefficients should be uncorrelated. In this case scalar quantization of transform coefficients followed by variable-length coding with rather low computational complexity can provide rate-distortion performance close to that theoretically achievable;
- transform should be orthonormal. If this property holds, then the average squared reconstruction error occurring in the input vector is equal to the average squared error introduced by quantizating the vector of the transform coefficients. If the transform is nonorthonormal, then sometimes even negligible quantization errors in some transform coefficient could lead to significant distortion of the input signal. Moreover, as will be shown below, any orthonormal transform preserves the achievable rate-distortion function $H(D)$ of the source obtained, assuming that the squared distortion measure is used;
- low computational complexity of computing transform coefficients. In particular, it is desirable to use the so-called separable two-dimensional (2D)-transforms for which the transform matrix can be factored into two matrices of one-dimensional (1D)-transforms.

Now we consider the main properties of transforms in more detail.

Let x be an input column vector of dimension N; then a linear transform of x can be expressed as follows

$$y = Tx$$

where $T = \{t_i\}, i = 1, 2, \ldots, N$ is an $N \times N$ transform matrix.

1. A transform is called *orthonormal* if

$$T^{-1} = T^* \tag{5.1}$$

where T is a matrix with complex elements and T^* denotes the adjoint matrix, i.e. the $N \times N$ matrix obtained from T by taking the transpose and then the complex conjugate of each entry. If T is a real matrix, then (5.1) reduces to the condition

$$T^{-1} = T^{\mathrm{T}}. \tag{5.2}$$

The condition (5.2) implies that the rows of the matrix $T = \{t_i\}$, $i = 1, \ldots, N$, satisfy the following equation

$$t_i t_j^{\mathrm{T}} = \delta_{ij}$$

where

$$\delta_{ij} = \begin{cases} 1, & \text{if } i = j \\ 0, & \text{if } i \neq j \end{cases}$$

is Kronecker's delta function.

The vectors t_i are called the orthonormal basis vectors of the linear transform T. For an orthonormal transform the following sequence of equalities holds

$$y = Tx$$
$$x = T^{-1}y$$
$$= T^{\mathrm{T}}y$$
$$= \sum_{i=1}^{N} y_i t_i^{\mathrm{T}} \tag{5.3}$$

that is, the input vector x can be represented as a weighted sum of basis vectors and the transform coefficients y_i act as weighting coefficients. If T is a complex matrix, then in (5.3) we should replace t_i by its complex conjugate. Harmonic functions are often used as basis functions. In this case the coefficients of the decomposition are interpreted as intensities of the corresponding harmonics. Transformations of this type are called *conversions to the frequency domain*.

The orthonormal transform preserves the signal energy, that is,

$$\sum_{i=1}^{N} x_i^2 = \sum_{j=1}^{N} |y_j|^2 .$$

This property of orthonormal transforms is known as the discrete Parseval's theorem and can be easily proved as follows

$$\sum_{i=1}^{N} x_i^2 = x^{\mathrm{T}}x = (T^{\mathrm{T}}y)^{\mathrm{T}}(T^{\mathrm{T}}y) = y^{\mathrm{T}}TT^{\mathrm{T}}y = y^{\mathrm{T}}TT^{-1}y = y^{\mathrm{T}}y = \sum_{j=1}^{N} |y_j|^2 .$$

2. Noncorrelatedness of the transform coefficients.

This property implies that the transform coefficients y_j, $j = 1, 2, \ldots, N$, satisfy the following condition

$$E\{(y_i - E\{y_i\})(y_j - E\{y_j\})\} = \alpha_i \delta_{ij}, \forall i, j \tag{5.4}$$

where α_i is the variance of the coefficient y_i and $E\{\cdot\}$ denotes the mathematical expectation. In the sequel, we assume for simplicity that $E\{x\} = E\{y\} = 0$.

3. Localization of most part of the signal energy in a small number of transform coefficients.

Assume that the transform coefficients $y_1, \ldots, y_N$ are sorted in order of decreasing variances, that is, $\alpha_1 \geq \alpha_2 \geq, \ldots, \geq \alpha_N$. We assume also that only first pN, $0 < p < 1$, coefficients are transmitted. The receiver uses a truncated vector-column $\hat{y} = (y_1, \ldots, y_{pN}, 0, \ldots, 0)^{\mathrm{T}}$ in order to form the reconstructed values $\hat{x} = T^{-1}\hat{y}$.

We will consider only orthonormal transforms, that is, $\hat{x} = T^{\mathrm{T}}\hat{y}$. The MSE occurring when we replace the original vector x by the reconstructed vector $\hat{x}$ is determined as follows

$$\mathrm{E}\left\{\frac{1}{N}\sum_{i=1}^{N}(x_i - \hat{x}_i)^2\right\} = \frac{1}{N}\mathrm{E}\left\{(x - \hat{x})^{\mathrm{T}}(x - \hat{x})\right\}$$

$$= \frac{1}{N}\mathrm{E}\left\{\left(y^{\mathrm{T}}T - \hat{y}^{\mathrm{T}}T\right)\left(T^{\mathrm{T}}y - T^{\mathrm{T}}\hat{y}\right)\right\}$$

$$= \frac{1}{N}\mathrm{E}\left\{\sum_{j=1}^{N}(y_j - \hat{y}_j)^2\right\}$$

$$= \frac{1}{N}\mathrm{E}\left\{\sum_{j=pN+1}^{N} y_j^2\right\} = \frac{1}{N}\sum_{j=pN+1}^{N}\alpha_j. \qquad (5.5)$$

Thus, we would like to find the orthonormal transform which minimizes the error (5.5).

We showed that when using an orthonormal transform the MSE occurring in the transform domain coincides with the MSE occurring in the domain of the original signal. Moreover, any orthonormal transform preserves the achievable rate-distortion function $H(D)$ of the source obtained under assumption that the squared distortion measure is used. Let X^N be a set of random input vectors $x = (x_1, x_2, \ldots, x_N)$. Denote as Y^N a set of vectors of the transform coefficients $y = (y_1, y_2, \ldots, y_N)$, $y = Tx$. Assume that the vector y is approximated with the MSE $D_n \leq D$ by a vector $\hat{y} = (\hat{y}_1, \hat{y}_2, \ldots, \hat{y}_N)$ from an approximating set $\hat{Y}^N$. Then the vector $\hat{y}$ is transformed into the vector $\hat{x}$ from the set $\hat{X}^N$, that is, $\hat{x} = T^{\mathrm{T}}\hat{y}$. Since an orthonormal transform is invertible, we obtain

$$I(Y^N, \hat{Y}^N) = I(X^N, \hat{X}^N).$$

It follows from (5.5) that if the MSE occurring when we approximate a vector from Y^N by a vector from $\hat{Y}^N$ is less than or equal to D, then a vector from X^N is also approximated by a vector from $\hat{X}^N$ with the MSE less than or equal to D. If Y^N is approximated by $\hat{Y}^N$ in such a way that

$$\frac{1}{N}I(Y^N, \hat{Y}^N)$$

takes its minimal value, then X^N is also approximated by $\hat{X}^N$ in such a way that

$$\frac{1}{N}I(X^N, \hat{X}^N)$$

takes its minimal value. Thus, the achievable rate-distortion function $H_X(D)$ of X^N coincides with the achievable rate-distortion function $H_Y(D)$ of Y^N. This property of an orthonormal transform guarantees that if transform coefficients are independent, even applying scalar quantization to them, we will not lose much compared to the achievable rate-distortion function of the given source.

4. Low computational complexity.

   In order to transform a 2D-input signal described by a matrix X of size $N \times N$, we have to apply a 2D-transform to it. For general nonseparable transforms we rearrange the input matrix X into a vector x with N^2 components and multiply it by the transform matrix T_2 of size $N^2 \times N^2$. The output $N \times N$ matrix of transform coefficients arranged as a column vector y of dimension N^2 is computed as

   $$y = T_2 x.$$

We call a linear transform separable if the $N^2 \times N^2$ transform matrix T_2 can be represented as the Kronecker product of two $N \times N$ matrices T of a 1D-transform. In this case the $N \times N$ matrix Y of N^2 transform coefficients can be computed by the formula

$$Y = T X T^{\mathrm{T}}.$$

In other words, first we transform each of the N columns (rows) of the matrix X by the 1D-transform T (we compute $Z = TX$ or $Z = XT^{\mathrm{T}}$) and then each of the N rows (columns) of the obtained matrix Z of the N^2 transform coefficients is transformed by using the same 1D-transform (we compute $Y = ZT^{\mathrm{T}}$ or $Y = TZ$). This means that the transform is implemented as two matrix multiplications of size $N \times N$ instead of one multiplication of a vector of size $N^2 \times 1$ with a matrix of size $N^2 \times N^2$ and the computational complexity is reduced to $2N^3$ operations instead of N^4.

In the following sections we analyze different transforms in terms of the above-mentioned requirements.

5.2   The Karhunen–Loeve transform

The so-called Karhunen–Loeve transform is the most efficient in terms of enumerated properties transforms. It is an orthonormal transform. Its coefficients satisfy condition (5.4); that is, they are uncorrelated (and hence independent if the input vector is Gaussian). Moreover, as we will show below, the Karhunen–Loeve transform minimizes (among all orthonormal transforms) the MSE (5.5) due to rejecting transform coefficients with small variances. In other words, this transform is optimum in terms of localization of the signal energy and it maximizes the number of transform coefficients which are insignificant and might be quantized to 0.

In order to find the matrix T_{KL} of the Karhunen–Loeve transform, we let x be the input vector-column of dimension N and $\mathrm{E}\{x\} = \mathrm{E}\{y\} = \mathbf{0}$. The covariance matrix of the vector x is expressed via covariance matrix of the vector of the transform coefficients y as follows

$$R = \mathrm{E}\left\{x x^{\mathrm{T}}\right\} = T_{\mathrm{KL}}^{\mathrm{T}} \mathrm{E}\left\{y y^{\mathrm{T}}\right\} T_{\mathrm{KL}}. \tag{5.6}$$

By multiplying both sides of (5.6) by T_{KL}^{T}, we obtain

$$RT_{\text{KL}}^{\text{T}} = T_{\text{KL}}^{\text{T}} \text{E}\left\{\boldsymbol{y}\boldsymbol{y}^{\text{T}}\right\}. \tag{5.7}$$

Since the transform coefficients are uncorrelated, the matrix $\text{E}\left\{\boldsymbol{y}\boldsymbol{y}^{\text{T}}\right\}$ is a diagonal matrix with variances λ_i, $i = 1, 2, \ldots, N$, along the principal diagonal. Taking into account this circumstance from (5.7), we obtain

$$Rt_{\text{KL}i}^{\text{T}} = \lambda_i t_{\text{KL}i}^{\text{T}}, i = 1, 2, \ldots, N.$$

Thus, we see that the basis vectors of the Karhunen–Loeve transform are eigenvectors of the covariance matrix R normalized to satisfy $t_{\text{KL}i}t_{\text{KL}j}^{\text{T}} = \delta_{ij}$. The variances $\{\lambda_i\}$ of the transform coefficients are eigenvalues of the matrix R. Since the matrix R is symmetric and a positive definite matrix (that is, $\boldsymbol{x}A\boldsymbol{x}^{\text{T}} > 0$ for any nonzero vector $\boldsymbol{x}$) the eigenvalues of this matrix are real and positive. As a result, the basis vectors and the transform coefficients of the Karhunen–Loeve transform are real. Equation (5.5) for the Karhunen–Loeve transform has the form

$$\frac{1}{N}\text{E}\left\{\sum_{j=pN+1}^{N} y_j^2\right\} = \frac{1}{N}\sum_{j=pN+1}^{N} \lambda_j.$$

Theorem 5.1 *Among all possible orthonormal transforms applied to stationary N-dimensional vectors, the Karhunen–Loeve transform minimizes the mean squared error (5.5) that occurs due to truncation.*

Proof. Let $U = \{\boldsymbol{u}_k\}$ be a matrix of some orthonormal transform. The vector-column of the transform coefficients $\boldsymbol{c} = (c_1, c_2, \ldots, c_N)^{\text{T}}$ is determined as

$$\boldsymbol{c} = U\boldsymbol{x}.$$

Then, the vector $\boldsymbol{x}$ can be decomposed over the basis vectors $\boldsymbol{u}_k$, $k = 1, 2, \ldots, N$, of this transform, that is,

$$\boldsymbol{x} = \sum_{k=1}^{N} c_k \boldsymbol{u}_k^{\text{T}}$$

where c_k are transform coefficients of the transform $U = \{\boldsymbol{u}_k\}$. Since $U = \{\boldsymbol{u}_k\}$ is an orthonormal transform, each coefficient c_k can be represented in the form

$$c_k = \boldsymbol{u}_k\boldsymbol{x}, k = 1, 2, \ldots, N.$$

The MSE that occurred because of truncating the vector $\boldsymbol{y}$ for this transform is determined as

$$\frac{1}{N}\sum_{k>pN} \text{E}\left\{|c_k|^2\right\} = \frac{1}{N}\sum_{k>pN} \boldsymbol{u}_k R \boldsymbol{u}_k^{\text{T}}. \tag{5.8}$$

Now let us decompose the basis vectors $\{u_k\}$ over the basis vectors $\{t_i\}$ of the Karhunen–Loeve transform. We obtain

$$u_k = \sum_{i=1}^{N} w_{ik} t_i$$

where

$$w_{ik} = u_k t_i^T \qquad (5.9)$$

and since $U = \{u_k\}$ is a unitary matrix

$$||u_k||^2 = \sum_{i=1}^{N} |w_{ik}|^2 = 1. \qquad (5.10)$$

On the other hand, the basis vectors of the Karhunen–Loeve transform $\{t_i\}$ in their turn can be represented in terms of the basis vectors $\{u_k\}$. It follows from (5.9) that $w_{ik}^T = t_i u_k^T$ and thereby

$$t_i = \sum_{k=1}^{N} w_{ik}^T u_k. \qquad (5.11)$$

Now we can write the following chain of equalities

$$u_k R u_k^T = \sum_{i=1}^{N} w_{ik} t_i R u_k^T = \sum_{i=1}^{N} w_{ik} \lambda_i t_i u_k^T = \sum_{i=1}^{N} |w_{ik}|^2 \lambda_i$$

$$= \lambda_{pN} + \sum_{i=1}^{N} |w_{ik}|^2 (\lambda_i - \lambda_{pN})$$

$$= \lambda_{pN} + \sum_{i=1}^{pN} |w_{ik}|^2 (\lambda_i - \lambda_{pN}) + \sum_{i>pN} |w_{ik}|^2 (\lambda_i - \lambda_{pN}). \qquad (5.12)$$

Taking into account that $\lambda_i \geq \lambda_{pN}$ for $i \leq pN$ from (5.12), we obtain

$$u_k R u_k^T \geq \lambda_{pN} + \sum_{i>pN} |w_{ik}|^2 (\lambda_i - \lambda_{pN}).$$

From (5.8) it follows that the MSE that occurred because of the truncation of the vector c of the transform coefficients of U is greater than or equal to

$$\frac{1}{N} \sum_{k>pN}^{N} \left(\lambda_{pN} - \sum_{i>pN}^{N} |w_{ik}|^2 (\lambda_{pN} - \lambda_i) \right) = \frac{(N - pN)}{N} \lambda_{pN}$$

$$- \frac{1}{N} \sum_{i>pN}^{N} (\lambda_{pN} - \lambda_i) \sum_{k>pN}^{N} |w_{ik}|^2.$$

From (5.10) it follows that

$$\sum_{k>pN}^{N} |w_{ik}|^2 \leq 1$$

and since $\lambda_{pN} - \lambda_i \geq 0$ for $i > pN$, the MSE, when we use the transform U, is greater than or equal to

$$(1 - p)\lambda_{pN} - \frac{1}{N} \sum_{i>pN}^{N} (\lambda_{pN} - \lambda_i) = \frac{1}{N} \sum_{i>pN}^{N} \lambda_i. \qquad (5.13)$$

$\square$

We found that the MSE that occurred due to the truncation of the vector of the transform coefficients of U is always greater than or equal to the MSE that occurred due to the truncation of the vector of the Karhunen–Loeve transform coefficients.

Thus, we proved that the Karhunen–Loeve transform is optimal in terms of localizing the signal energy. The main shortcoming of the Karhunen–Loeve transform is that its basis functions depend on the transformed signal. As a result, in order to reconstruct the synthesized signal from the quantized transform coefficients, we have to know the description of the basis functions. Hence we have to store and transmit not only quantized transform coefficients but also the basis functions which can require many more bits for storing than the quantized coefficients. Moreover, it is recognized that using basis functions that are known in advance usually makes it possible to exploit their algebraic properties to reduce the computational complexity of the transform.

5.3   The discrete Fourier transform

The discrete Fourier transform (DFT) is the counterpart of the continuous Fourier transform. It is defined for discrete-time signals. The transformed signal is also a discrete signal but in the frequency domain and it represents samples of the signal spectrum.

Let $x(nT_s)$ be a sequence of signal samples taken at time moments nT_s ($0 \leq n \leq N - 1$), then the DFT of $x(nT_s)$ is

$$y(k\Omega) = \sum_{n=0}^{N-1} x(nT_s)e^{-jkn\Omega T_s}, 0 \leq k \leq N - 1. \qquad (5.14)$$

The Inverse DFT (IDFT) is determined as follows

$$x(nT_s) = \frac{1}{N} \sum_{k=0}^{N-1} y(k\Omega)e^{jkn\Omega T_s}, 0 \leq n \leq N - 1 \qquad (5.15)$$

where $\Omega = 2\pi/(NT_s)$ is the distance between samples of the signal spectrum in the frequency domain. In other words, $y(k\Omega)$ represents N samples of the spectrum of $x(nT_s)$ taken with the period Ω over the frequency axis. Notice that (5.14) determines a periodical sequence of numbers with period N. Expressions (5.14) and (5.15) can be rewritten in the following form

$$y_k = \sum_{n=0}^{N-1} x_n W_N^{kn}, 0 \le k \le N - 1 \tag{5.16}$$

$$x_n = \frac{1}{N} \sum_{k=0}^{N-1} y_k W_N^{-kn}, 0 \le n \le N - 1 \tag{5.17}$$

where $W_N = e^{-j2\pi/N}$.

Henceforth we consider samples taken in time and frequency domains as vector components and write x_n instead of $x(n)$.

In matrix form, (5.14) and (5.15) can be written as

$$y = T_F x, \, x = T_F^{-1} y = \frac{1}{N} T_F^* y \tag{5.18}$$

where $y = (y_0, y_1, \ldots, y_{N-1})^T$, $x = (x_0, x_1, \ldots, x_{N-1})^T$, and

$$T_F = \begin{pmatrix} 1 & 1 & \ldots & 1 & 1 \\ 1 & W_N & \ldots & W_N^{N-2} & W_N^{N-1} \\ \ldots & \ldots & \ldots & \ldots & \ldots \\ 1 & W_N^{N-2} & \ldots & W_N^{(N-2)(N-2)} & W_N^{(N-2)(N-1)} \\ 1 & W_N^{N-1} & \ldots & W_N^{(N-1)(N-2)} & W_N^{(N-1)(N-1)} \end{pmatrix}.$$

Instead of (5.18) we will often use

$$y = DFT(x), \, x = IDFT(y).$$

Since $T_F^{-1} = \frac{1}{N} T_F^*$, DFT is an orthogonal transform. Notice that it is easy to normalize DFT in order to obtain an orthonormal transform. For this purpose we should use the factor $1/\sqrt{N}$ in both the forward and the inverse transforms instead of using the factor $1/N$ in the inverse transform, but the form of DFT presented above is used more often.

The 2D-DFT is defined as follows

$$Y_{k,l} = \sum_{n=0}^{N-1} \sum_{m=0}^{M-1} X_{n,m} e^{-j(kn\frac{2\pi}{N} + lm\frac{2\pi}{M})}, 0 \le k \le N - 1, \, 0 \le l \le M - 1$$

where $X_{n,m}$ is an element of the input matrix X and $Y_{k,l}$ is an element of the matrix of transform coefficients.

Since

$$Y_{k,l} = \sum_{m=0}^{M-1} \left\{ \sum_{n=0}^{N-1} X_{n,m} e^{-jkn\frac{2\pi}{N}} \right\} e^{-jlm\frac{2\pi}{M}}$$

$$= \sum_{n=0}^{N-1} \left\{ \sum_{m=0}^{M-1} X_{n,m} e^{-jlm\frac{2\pi}{M}} \right\} e^{-jkn\frac{2\pi}{N}}$$

the 2D-DFT can be split into two 1D-transforms; that is, DFT is a separable transform.

We have considered the DFT in terms of requirements related to transform coding techniques. Below, we also mention some other important properties of the DFT to which we will refer in the following sections:

- linearity
- circular convolution
- circular displacement.

Next, we consider these properties in more detail.

Property 5.1   *(Linearity) Let x_1 and x_2 be two input vectors of the same dimension, and a, b be arbitrary constants, then*

$$DFT\{ax_1 + bx_2\} = aDFT\{x_1\} + bDFT\{x_2\}.$$

Property 5.2   *(Circular convolution) Let $y_i = (y_{i,0}, y_{i,1}, \ldots, y_{i,N-1}) = DFT(x_i)$, $i = 1, 2$, where $x_i = (x_{i,0}, x_{i,1}, \ldots, x_{i,N-1})$. Let $a \circ b$ denote the componentwise product of two vectors a and b and $[a]_l$, $l = 0, 1, \ldots, N - 1$, be the lth component of vector a. Denote by $a \circledast b$ the circular convolution of vectors a and b, that is,*

$$[a \circledast b]_l = \sum_{n=0}^{N-1} a_n b_{((l-n))} = \sum_{n=0}^{N-1} b_n a_{((l-n))} \tag{5.19}$$

where $((l)) = l \mod N$. The circular convolution is a counterpart to the ordinary convolution.

It is easy to see that the circular convolution (5.19) of two sequences differs from the ordinary convolution of two sequences, since all indices in (5.19) are computed modulo N, where N is the length of the sequences.

Theorem 5.2

$$DFT(x_1 \circ x_2) = DFT(x_1) \circledast DFT(x_2)$$

and

$$DFT(x_1 \circledast x_2) = DFT(x_1) \circ DFT(x_2).$$

Proof. According to the definition $IDFT(y_1 \circ y_2)$ is

$$[IDFT(y_1 \circ y_2)]_l = \frac{1}{N} \sum_{k=0}^{N-1} y_{1k} y_{2k} e^{j\frac{2\pi}{N}kl}. \tag{5.20}$$

By inserting the definitions for y_1 and y_2 into (5.20), we obtain

$$\left[IDFT\left(y_1 \circ y_2\right)\right]_l = \frac{1}{N} \sum_{k=0}^{N-1} \left\{ \sum_{n=0}^{N-1} x_{1n} e^{-j\frac{2\pi}{N}kn} \right\}$$

$$\times \left\{ \sum_{m=0}^{N-1} x_{2m} e^{-j\frac{2\pi}{N}km} \right\} e^{j\frac{2\pi}{N}kl}.$$

Changing the order of summation, we obtain

$$\left[IDFT\left(y_1 \circ y_2\right)\right]_l = \frac{1}{N} \sum_{n=0}^{N-1} \sum_{m=0}^{N-1} x_{1n} x_{2m} \left\{ \sum_{k=0}^{N-1} e^{j\frac{2\pi k}{N}(l-n-m)} \right\}.$$

The sum in the brackets is equal to 0 for all m and n except when m and n satisfy the equality $m = (l - n) \mod N$ for which the sum in brackets is equal to N. Thus, (5.20) can be reduced to expression (5.19). We obtain that the product of the DFTs is the DFT of the *circular or periodical convolution*. $\qquad \square$

Example 5.1 Let $x = (2, 3, 1, 2)^T$ and $y = (1, 2, 4, 3)^T$, $N = 4$.
The DFT matrix is

$$T_F = \begin{pmatrix} 1 & 1 & 1 & 1 \\ 1 & -j & -1 & j \\ 1 & -1 & 1 & -1 \\ 1 & j & -1 & -j \end{pmatrix} \qquad (5.21)$$

$$DFT(x) = (8, 1 - j, -2, 1 + j)^T$$
$$DFT(y) = (10, -3 + j, 0, -3 - j)^T$$
$$DFT(x) \circ DFT(y) = (80, -2 + 4j, 0, -2 - 4j)^T.$$

The matrix of the inverse transform is

$$T_F^{-1} = \frac{1}{4} \begin{pmatrix} 1 & 1 & 1 & 1 \\ 1 & j & -1 & -j \\ 1 & -1 & 1 & -1 \\ 1 & -j & -1 & j \end{pmatrix} \qquad (5.22)$$

$$x \circledast y = IDFT(80, -2 + 4j, 0, -2 - 4j)^T$$
$$= (19, 18, 21, 22)^T.$$

By applying (5.19) directly, we obtain the same coefficients v_l of the circular convolution

$$v_0 = y_0 x_0 + y_1 x_3 + y_2 x_2 + y_3 x_1 = 19$$

$$v_1 = y_0 x_1 + y_1 x_0 + y_2 x_3 + y_3 x_2 = 18$$

$$v_2 = y_0 x_2 + y_1 x_1 + y_2 x_0 + y_3 x_3 = 21$$

$$v_3 = y_0 x_3 + y_1 x_2 + y_2 x_1 + y_3 x_0 = 22.$$

The computation of the circular convolution can be interpreted as follows. In order to compute the circular convolution of the vectors x and y, we periodically continue x and consider the infinite sequence

$$x^{\text{ext}} = (\ldots, x, x, x, \ldots) = (\ldots, 2, 3, 1, 2, 2, 3, 1, 2, 2, 3, 1, 2, \ldots).$$

The periodical sequence x^{ext} is reversed in time and multiplied by the corresponding term in the sequence y. Then the reversed in time sequence x^{ext} is shifted by one sample to the right and again multiplied by y, generating the next coefficient v_l.

Reversed x^{ext}	2	1	3	2	2	1	3	2	2	1	3	2
Sequence y				1	2	4	3					
x^{ext} shifted by 1	2	2	1	3	2	2	1	3	2	2	1	3
x^{ext} shifted by 2	3	2	2	1	3	2	2	1	3	2	2	1
x^{ext} shifted by 3	1	3	2	2	1	3	2	2	1	3	2	2

Property 5.3 *(Circular displacement) Let*

$$x^{(l)} = (x_{((0-l))}, x_{((1-l))}, \ldots, x_{((N-1-l))})$$
$$= (x_{N-l}, x_{N-l+1}, \ldots, x_{l+1})$$

then we have the following

Theorem 5.3

$$\left[DFT \left(x^{(l)} \right) \right]_k = [DFT(x)]_k \, W_N^{lk}.$$

Proof. Letting y be the vector

$$(\underbrace{0, 0, \ldots, 1}_{l}, 0, \ldots, 0)$$

from Theorem (5.2), we obtain the statement of the theorem.   □

The displacement of l samples from the end to the beginning of the sequence x is equivalent to multiplying its DFT by $\exp\{-j2\pi lk/N\}$.

In general, the DFT and the IDFT require approximately N^2 additions and N^2 multiplications of complex numbers. Direct computation of the DFT is inefficient because it does not exploit the symmetry of the phase factor W_N; that is, $W_N^{k+N/2} = -W_N^k$ and its periodicity $W_N^{k+N} = W_N^k$. There exist so-called *fast algorithms* which reduce the computational complexity of DFT to $N \log_2 N$ operations. There are two classes of fast Fourier transform (FFT) algorithms. They are called *decimation-in-time* and *decimation-in-frequency* algorithms. Each class contains many modifications.

Here we consider the basic ideas used in decimation-in-time algorithms. Assume that $N = 2^p$. Notice that the DFT of length N is often referred to as the *N-point* DFT. We split the input sequence x of length N into two sequences x_1 and x_2 of length $N/2$ each. They contain the even-numbered and the odd-numbered samples, respectively, that is,

$$
\begin{aligned}
x_1 &= (x_{1,0}, x_{1,1}, \ldots, x_{1,N/2-1}) \\
&= (x_0, x_2, \ldots, x_{N-2}) \\
x_2 &= (x_{2,0}, x_{2,1}, \ldots, x_{2,N/2-1}) \\
&= (x_1, x_3, \ldots, x_{N-1}).
\end{aligned}
$$

The DFTs for the sequences x_1 and x_2 are

$$
g_k = \sum_{n=0}^{N/2-1} x_{1,n} W_N^{2nk} \tag{5.23}
$$

$$
h_k = \sum_{n=0}^{N/2-1} x_{2,n} W_N^{2nk} \tag{5.24}
$$

where $k = 0, 1, \ldots, N/2 - 1$.

Let $y = DFT(x)$, then we can express y via (5.23) and (5.24) as

$$
y_k = \sum_{n=0}^{N/2-1} x_{1,n} W_N^{2nk} + x_{2,n} W_N^{(2n+1)k} = g_k + h_k W_N^k
$$

where $k = 0, 1, \ldots, N/2$. Taking into account that $g_{k+N/2} = g_k$, $h_{k+N/2} = h_k$, and $W_N^{k+N/2} = -W_N^k$, we obtain

$$
y_k = \begin{cases} g_k + h_k W_N^k, & \text{if } k = 0, 1, \ldots, N/2 - 1 \\ g_k - h_k W_N^{k-N/2}, & \text{if } k = N/2, N/2 + 1, \ldots, N - 1. \end{cases} \tag{5.25}
$$

The direct computation of $DFT(x_1)$ and $DFT(x_2)$ requires $(N/2)^2$ complex multiplications. Additionally, $N/2$ multiplications are required in order to compute $W_N^k [DFT(x_2)]_k$. In total, $N^2/2 + N/2$ complex operations are required instead of N^2 operations for the direct computation. It is easy to see that the number of operations is reduced approximately by a factor of two. Figure 5.2 illustrates the first step of the time-decimation algorithm for $N = 8$. The 8-point DFT is reduced to two 4-point DFTs. We can split each of the sequences x_1 and x_2 into two subsequences of length $N/4$ and reduce the computation of the $N/2$-point DFT to the computation of two $N/4$-point DFTs. We continue the decimation of the input sequence until we come to 2-point

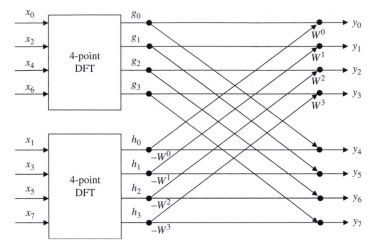

Figure 5.2   First step of computation of 8-point DFT

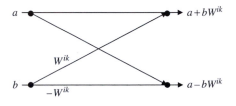

Figure 5.3   Two-point DFT

DFTs. In Fig. 5.3 the diagram of a 2-point DFT is shown. If $N = 8$, we perform four 2-point DFTs with $a = x_0$, $b = x_4$; $a = x_2$, $b = x_6$; $a = x_1$, $b = x_5$; $a = x_3$, $b = x_7$, respectively. If $N = 2^p$, then the input sequence can be decimated $p = \log_2 N$ times. In Fig. 5.4 three steps of the decimation-in-time algorithm for $N = 8$ are illustrated. It can be shown that the total number of operations is proportional to $N \log_2 N$.

5.4   The discrete cosine transform

Coding techniques using the DCT are the basis of all modern standards of image and video compression. Consider a vector $\boldsymbol{x} = (x_0, x_1, \ldots, x_{N-1})^{\mathrm{T}}$ with covariance matrix in the form

$$\Lambda = \sigma^2 \begin{pmatrix} 1 & \rho & \rho^2 & \ldots & \rho^{N-1} \\ \rho & 1 & \rho & \ldots & \rho^{N-2} \\ \rho^2 & \rho & 1 & \ldots & \rho^{N-3} \\ \ldots & \ldots & \ldots & \ldots & \ldots \\ \rho^{N-1} & \rho^{N-2} & \rho^{N-3} & \ldots & 1 \end{pmatrix}$$

where $\rho = \mathrm{E}\{(x_i - \mathrm{E}\{x_i\})(x_{i+1} - \mathrm{E}\{x_{i+1}\})\}$ is the correlation coefficient between two neighboring samples and σ^2 denotes the variance of the samples. It is

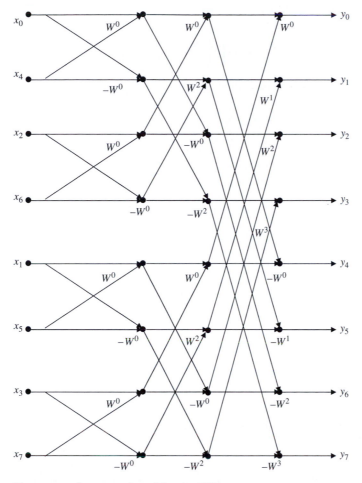

Figure 5.4 Three steps of computation of 8-point DFT

known (Jayant and Noll 1984) that if $\rho \to 1$, then the eigenvectors of this matrix are the sampled sine waves with phases depending on the eigenvalues which are uniformly distributed in the interval $(0, \pi)$. As a result, for images with highly correlated samples ($\rho > 0.7$) the efficiency of the DCT in terms of localization of the signal energy is close to the efficiency of the Karhunen–Loeve transform. On the other hand, the DCT represents the orthonormal separable transform which does not depend on the transformed image and thus its computational complexity is rather low. These properties of the DCT explain the fact that this transform was chosen as the standard solution for video compression.

The DCT decomposes the signal using a set of N different cosine waveforms sampled at N points. There are two commonly used types of DCT. The so-called DCT-II is used in JPEG and MPEG standards. It is defined as

$$y_k = \frac{c_k \sqrt{2}}{\sqrt{N}} \sum_{n=0}^{N-1} x_n \cos\left(\frac{(n+1/2)k\pi}{N}\right)$$

(5.26)

where

$$
c_k = \begin{cases} 1/\sqrt{2}, & \text{if } k = 0 \\ 1, & \text{if } k \neq 0. \end{cases}
$$

The inverse transform can be written as

$$
x_n = \sqrt{\frac{2}{N}} \sum_{k=0}^{N-1} y_k c_k \cos\left(\frac{(n+1/2)k\pi}{N}\right). \tag{5.27}
$$

Notice that factor c_k is introduced to normalize the transform.
We can write (5.26) and (5.27) in the matrix form as

$$
\boldsymbol{y} = T_{\mathrm{D}} \boldsymbol{x}, \; \boldsymbol{x} = T_{\mathrm{D}}^{-1} \boldsymbol{y} = T_{\mathrm{D}}^{\mathrm{T}} \boldsymbol{y} \tag{5.28}
$$

where $\boldsymbol{y} = (y_0, y_1, \ldots, y_{N-1})^{\mathrm{T}}$, $\boldsymbol{x} = (x_0, x_1, \ldots, x_{N-1})^{\mathrm{T}}$,

$$
T_{\mathrm{D}} = \sqrt{\frac{2}{N}} \begin{pmatrix} 1/\sqrt{2} & 1/\sqrt{2} & \cdots & 1/\sqrt{2} \\ \cos\left(\frac{\pi}{2N}\right) & \cos\left(\frac{3\pi}{2N}\right) & \cdots & \cos\left(\frac{(2N-1)\pi}{2N}\right) \\ \cdots & \cdots & \cdots & \cdots \\ \cos\left(\frac{(N-2)\pi}{2N}\right) & \cos\left(\frac{3(N-2)\pi}{2N}\right) & \cdots & \cos\left(\frac{(2N-1)(N-2)\pi}{2N}\right) \\ \cos\left(\frac{(N-1)\pi}{2N}\right) & \cos\left(\frac{3(N-1)\pi}{2N}\right) & \cdots & \cos\left(\frac{(2N-1)(N-1)\pi}{2N}\right) \end{pmatrix}.
$$

It follows from (5.28) that $T_{\mathrm{D}}^{-1} = T_{\mathrm{D}}^{\mathrm{T}}$, that is, the DCT is an orthonormal transform and the cosine basis functions are orthogonal.

The 2D-DCT is determined as

$$
Y_{k,l} = 2 \frac{c_k}{\sqrt{N}} \frac{c_l}{\sqrt{M}} \sum_{n=0}^{N-1} \sum_{m=0}^{M-1} X_{n,m} \cos\left(\frac{(n+1/2)k\pi}{N}\right)
$$

$$
\times \cos\left(\frac{(m+1/2)l\pi}{M}\right). \tag{5.29}
$$

The inverse 2D-DCT can be written as

$$
X_{n,m} = \frac{2}{\sqrt{MN}} \sum_{k=0}^{N-1} \sum_{l=0}^{M-1} Y_{k,l} c_k c_l \cos\left(\frac{(n+1/2)k\pi}{N}\right)
$$

$$
\times \cos\left(\frac{(m+1/2)l\pi}{M}\right)
$$

where $X_{n,m}$ is an element of the input matrix X; $Y_{k,l}$ is an element of the matrix Y of transform coefficients.

It is easy to see that (5.29) can be represented in the form

$$
Y_{k,l} = \frac{\sqrt{2}c_k}{\sqrt{N}} \sum_{n=0}^{N-1} Z_{n,l} \cos\left(\frac{(n+1/2)k\pi}{N}\right)
$$

where $Z = \{Z_{n,e}\}$ is the output of the 1D-DCT performed over the rows of the matrix X or, in other words, it is a separable transform.

The DCT-IV is used in the MPEG-audio standard as the basis of the so-called modified DCT (a kind of overlapped transform). This transform is defined as

$$y_k = \sqrt{\frac{2}{N}} \sum_{n=0}^{N-1} x_n \cos\left(\frac{(n+1/2)(k+1/2)\pi}{N}\right). \tag{5.30}$$

The inverse DCT-IV has the same form,

$$x_n = \sqrt{\frac{2}{N}} \sum_{k=0}^{N-1} y_k \cos\left(\frac{(n+1/2)(k+1/2)\pi}{N}\right).$$

This transform is also orthonormal and separable. There is no constant among its basis functions. Notice that one of the good properties of DCT-IV is that its matrix is symmetric. It needs no c_k to insert $\sqrt{2}$ into a zero-frequency term. Both transforms can be reduced to the DFT and thus computed using the FFT.

Let us consider how DCT-II can be implemented via DFT. One possible approach is the following. We start with reordering x. Let v be an auxiliary vector with components

$$v_n = x_{2n}$$

and

$$v_{N-n-1} = x_{2n+1} \text{ for } n = 0, 1, \ldots, \frac{N}{2} - 1.$$

Now we take the DFT of v. The DCT-II coefficients are

$$
\begin{aligned}
y_k &= \cos\left(\frac{\pi k}{2N}\right) Re\{w_k\} + \sin\left(\frac{\pi k}{2N}\right) Im\{w_k\} \\
&= \sum_{n=0}^{N-1} v_n \left(\cos\left(\frac{2\pi nk}{N}\right)\cos\left(\frac{\pi k}{2N}\right) - \sin\left(\frac{2\pi nk}{N}\right)\sin\left(\frac{\pi}{2N}\right)\right) \\
&= \sum_{n=0}^{N-1} v_n \cos\left(\frac{(4n+1)\pi k}{2N}\right)
\end{aligned}
\tag{5.31}
$$

where $k = 0, 1, \ldots, N - 1$ and $w = DFT(v)$.

Let us check that (5.31) and (5.26) give the same DCT-II coefficients. It is evident that even-numbered samples $x_{2n}, n = 0, 1, \ldots, N/2 - 1$ in (5.31) we multiply by

$$\cos\left(\frac{(4n+1)\pi k}{2N}\right)$$

that coincides with the required coefficient

$$\cos\left(\frac{(2n+1/2)\pi k}{N}\right).$$

Odd-numbered samples x_{2n+1} we multiply by

$$\cos\left(\frac{(4(N-n-1)+1)\pi k}{2N}\right) = \cos\left(\frac{(4N-(4n+3))\pi k}{2N}\right)$$
$$= \cos\left(\frac{(2n+1+1/2)\pi k}{N}\right).$$

Now we consider how DCT-IV can be implemented via the DFT of length $N/2$. Reordering the input sequence x_n, $n = 0, 1, \ldots, N-1$, we obtain an auxiliary sequence of $N/2$ complex numbers

$$v_n = (x_{2n} + j x_{N-1-2n}) \exp\left(-j\frac{\pi}{N}\left(n+\frac{1}{4}\right)\right).$$

For the sequence v_n, $n = 0, 1, \ldots, N-1$ we compute the DFT of length $N/2$, that is,

$$w_k = \sum_{n=0}^{N/2-1} v_n e^{-j\frac{4nk\pi}{N}}.$$

The DCT-IV coefficients y_k, $k = 0, 1, \ldots, N-1$, can be obtained by the formulas

$$y_{2k} = Re\left\{c_k \exp\left(-jk\frac{\pi}{N}\right)\right\}$$

and

$$y_{N-1-2k} = -Im\left\{c_k \exp\left(-jk\frac{\pi}{N}\right)\right\}$$

where $c_k = \sqrt{\frac{2}{N}} w_k$.

Problems

5.1 Show that the matrix $T = \{t_k\} = \{t_{ki}\}$ defined by

$$t_{ki} = \frac{1}{\sqrt{N}} \exp(-j\frac{2\pi}{N}ki), k, i = 0, 1, \ldots, N-1$$

determines an orthonormal transform.

5.2 Let T denote the transform matrix of the DFT of size 4×4. Write down the explicit form of the transform matrix. Show that the transform is orthogonal.

5.3 Let T denote the transform matrix of the DCT-II of size 4×4. Write down the explicit form of the transform matrix. Show that the transform is orthogonal.

5.4 Let T_2 denote the 4×4 transform matrix of the 2D-DCT-II which can be applied to the input blocks of size 2×2. Write down the explicit form of the transform matrix. Show that the transform is separable.

5.5 Apply the DCT-II and DCT-IV to the vectors $x = (1, \ 1, \ 1, \ 1)^{\mathrm{T}}$ and $x = (1, -1, 1, -1)^{\mathrm{T}}$. Compare the obtained results. For which of the transforms is the localization of the signal energy better? Explain the obtained results.

5.6 Apply the DCT-II and the DCT-IV to the vectors $x = (1, 0, 0, 0)^{\mathrm{T}}$ and $x = (0, 1, 0, 0)^{\mathrm{T}}$. Compare the obtained results with the solution of Problem 5.5. For which of the vectors is the localization of the signal energy better? Explain the obtained results.

5.7 Check that reordering v_n as $v_n = x_{2n}$ and $v_{N-n-1} = x_{2n+1}$ for $n = 0, 1, \ldots, N/2 - 1$ transforms (5.26) into

$$y_k = \frac{\sqrt{2}c_k}{\sqrt{N}} \sum_{n=0}^{N-1} v_n \cos \frac{\pi (4n + 1)k}{2N}.$$

5.8 Let the covariance matrix of the input vector be equal to

$$R = \begin{pmatrix} 1 & \rho \\ \rho & 1 \end{pmatrix}.$$

Find the transform matrix of the Karhunen–Loeve transform.

5.9 Assume that the output vector $x = (x_1, x_2)$ of the Gaussian source described by the covariance matrix

$$R = \begin{pmatrix} 1 & \rho \\ \rho & 1 \end{pmatrix}$$

is transformed by the Karhunen–Loeve transform into the vector $y = (y_1, y_2)$. Assume also that the components of y are sorted in the order of decreasing variances. Truncate y by setting the component with the smallest variance to zero. Compute the squared error occurring due to this truncation.

6 Filter banks and wavelet filtering

We have considered two linear transforms (DFT and DCT) which are based on decomposition of an input signal over a system of orthogonal harmonic functions. The main shortcoming of these transforms is that their basis functions are uniformly distributed over the frequency axis. It means that all frequency components of the input signal are considered as equally important in terms of recovering the original signal from the set of transform coefficients. On the other hand, from the signal reconstruction point of view it is more important to preserve the high quality of low-frequency components of the signal than to preserve its high-frequency components. Thus, the resolution of the system of basis functions should be nonuniform over the frequency axis. The problem of constructing such a transform is solved by using filter banks. A traditional *subband coding* is a technique based on using filters of equal or roughly equal bandwidths. One of the most efficient transforms is based on wavelet filter banks and is called *wavelet filtering* (Strang and Nguyen 1996). Wavelet filter banks can be regarded as the subband coding with logarithmically varying filter bandwidths where the filters satisfy certain properties.

6.1 Linear filtering as a linear transform

It is well known that the output sequence $y(n)$ of the discrete-time filter with the pulse response $h(n)$ is the convolution of the input sequence $x(n)$ with $h(n)$; that is,

$$y(n) = \sum_{k=0}^{n} h(k)x(n-k) = \sum_{k=0}^{n} h(n-k)x(k). \tag{6.1}$$

The convolution (6.1) can be rewritten in matrix form as follows

$$y = Tx$$

where

$$T = \begin{pmatrix} h(0) & 0 & 0 & 0 & \dots \\ h(1) & h(0) & 0 & 0 & \dots \\ h(2) & h(1) & h(0) & 0 & \dots \\ h(3) & h(2) & h(1) & h(0) & \dots \\ \dots & h(3) & h(2) & h(1) & \dots \\ \dots & \dots & \dots & \dots & \dots \end{pmatrix}. \tag{6.2}$$

In other words, filtering is equivalent to the linear transform described by the constant-diagonal matrix T.

Assume that the input sequence $x(n)$ is an infinite periodical sequence of period N; that is,

$$x = (\ldots, x(0), x(1), \ldots, x(N-1), x(0), x(1), \ldots, x(N-1), \ldots)^\mathrm{T}$$

and consider a FIR filter with the pulse response $h(n)$ of length l. Then the output sequence $y(n)$ is also periodical with the same period N,

$$y = (\ldots, y(0), y(1), \ldots, y(N-1), y(0), y(1), \ldots, y(N-1), \ldots)^\mathrm{T}$$

where

$$y(m) = \sum_{k=0}^{l-1} h(k)x((m-k) \quad \mathrm{mod}\ N), m = 0, 1, \ldots, N-1 \tag{6.3}$$

is the circular convolution of one period of x, i.e. $x(0), x(1), \ldots, x(N-1)$ and the pulse response of the filter.

Our goal is to construct a linear transform based on linear filtering which converts a time-limited input sequence of length N into an output sequence of length N and can be described by the $N \times N$ transform matrix T. However, it is well known that any linear filtering delays the input sequence by $l - 1$ where l denotes the length of the filter pulse response. As a result, the corresponding linear transform converts an input sequence of length N into an output sequence of length $N + l - 1$. In order to avoid this problem we can do the following. If the input sequence $x(n)$ is a time-limited sequence of length N, we can always periodically extend it to obtain the infinite input vector x. We start filtering this infinite input sequence with the sample $x(N - l + 1)$ and stop after $N + l - 1$ steps. Then we take the samples $y(l - 1), y(l), \ldots, y(N + l - 2)$ of the output sequence $y(n)$. The last operation can be implemented by windowing $y(n)$ with a rectangular window of length N. In matrix form the corresponding transform can be written as

$$y = Tx$$

where $x = (x(0), x(1), \ldots, x(N-1))^\mathrm{T}$, $y = (y(0), y(1), \ldots, y(N-1))^\mathrm{T}$, and the $N \times N$ transform matrix T has the form

$$T = \begin{pmatrix} h(0) & 0 & \ldots & 0 & h(l-1) & \ldots & h(1) \\ h(1) & h(0) & 0 & \ldots & 0 & h(l-1) & \ldots \\ \ldots & \ldots & \ldots & \ldots & \ldots & \ldots & \ldots \\ 0 & \ldots & 0 & h(l-1) & h(l-2) & \ldots & h(0) \end{pmatrix}. \tag{6.4}$$

It is easy to see that (6.4) describes a circular convolution (6.3) of the input sequence and the pulse response of the filter. Next in this chapter, we consider some examples from (Strang and Nguyen 1996).

6.2   "Moving average" filtering as a linear transform

Consider the simplest lowpass FIR filter described by

$$y(n) = \frac{1}{2}x(n) + \frac{1}{2}x(n-1). \tag{6.5}$$

It is called *moving average* because its output is the average of the current sample $x(n)$ and the previous sample $x(n-1)$. The filter coefficients are $h(0) = 1/2$, $h(1) = 1/2$ and the length of its pulse response is equal to 2. Assume that the input sequence $x(n)$ has the length N. Then the corresponding $N \times N$ transform matrix has the form

$$T = \begin{pmatrix} \frac{1}{2} & 0 & 0 & \cdots & 0 & \frac{1}{2} \\ \frac{1}{2} & \frac{1}{2} & 0 & 0 & \cdots & 0 \\ 0 & \frac{1}{2} & \frac{1}{2} & 0 & \cdots & 0 \\ \cdots & \cdots & \cdots & \cdots & \cdots & \cdots \\ 0 & 0 & \cdots & 0 & \frac{1}{2} & \frac{1}{2} \end{pmatrix}. \tag{6.6}$$

The frequency response $H(e^{j\omega T_s})$ of the filter described by (6.5) can be written as

$$H(e^{j\omega T_s}) = \sum_n h(n)e^{-j\omega n T_s} = \frac{1}{2} + \frac{1}{2}e^{-j\omega T_s}. \tag{6.7}$$

By factoring out $e^{-j\omega T_s/2}$, we obtain

$$H(e^{j\omega T_s}) = \left(\frac{e^{-j\omega T_s} + e^{j\omega T_s}}{2} \right) e^{-j\omega T_s/2} = \cos\left(\frac{\omega T_s}{2} \right) e^{-j\omega T_s/2}. \tag{6.8}$$

It follows from (6.8) that the amplitude and the phase are determined as follows:

$$H(\omega) = \cos\left(\frac{\omega T_s}{2} \right), \quad H(\alpha) = \cos(\pi\alpha)$$

and

$$\varphi(\omega) = -\frac{\omega T_s}{2}, \quad \varphi(\alpha) = -\pi\alpha$$

where

$$\alpha = \omega/\omega_s.$$

In Fig. 6.1 the plots of the amplitude $H(\alpha)$ and the phase $\varphi(\alpha)$ are shown. It is easy to see that the moving average is a lowpass filter. It preserves the lowest frequency component which is the Direct Current (DC) term since $\cos\alpha = 1$ if $\alpha = 0$. The frequency response is small or zero near the highest discrete-time frequency $\alpha = 1/2$. Another important property of this filter is that it has *linear phase*. This property means that the filter does not distort the phase of the output signal $y(n)$ but only delays it with respect to the input signal $x(n)$.

In order to use linear filtering as a preprocessing step in a multimedia compression system, we need an invertible transform which will not introduce any error in the input

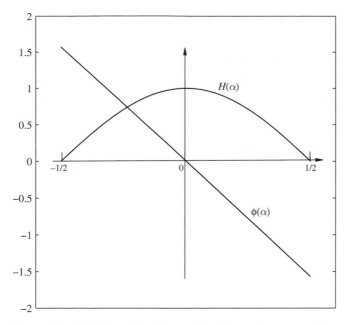

Figure 6.1   The amplitude $H(\alpha)$ and the phase $\varphi(\alpha)$ of the moving average lowpass filter

signal. However, it is easy to see that the averaging filter is not invertible. The frequency response of an invertible filter must have $H(\alpha) \neq 0$ at all frequencies since a linear filter can recover only zero input from zero output. For a moving average filter this requirement is not satisfied since $H(1/2) = 0$. The alternating signal

$$x(n) = \ldots, 1, -1, 1, -1 \ldots$$

is transformed to the all-zero output signal.

Notice that when the inverse filter for the filter described by $H(e^{j2\pi\alpha})$ exists it has the frequency response $1/H(e^{j2\pi\alpha})$. Formally we can write the frequency response for the inversion of the moving average

$$\frac{1}{\frac{1}{2} + \frac{1}{2}e^{-j2\pi\alpha}} = 2\left(1 - e^{-j2\pi\alpha} + e^{-j4\pi\alpha} - e^{-j6\pi\alpha} + \cdots\right) \qquad (6.9)$$

but this filter is unstable. The series expansion in (6.9) does not converge if $\alpha = 1/2$ since $1 + e^{-j\pi} = 0$:

$$\frac{1}{0} = \frac{1}{\frac{1}{2} + \frac{1}{2}e^{-j\pi}} = 2(1 + 1 + 1 + 1 + \cdots).$$

It is not always the case that a lowpass discrete-time FIR filter has no inverse. For example, the lowpass filter with coefficients $h(0) = 1/2$ and $h(1) = 1/4$ is invertible. Its frequency response $1/2 + e^{-j2\pi\alpha}/4$ is never equal to zero. The response at $\alpha = 0$ is $1/2 + 1/4 = 3/4$. The response at $\alpha = 1/2$ is $1/2 - 1/4 = 1/4$. The frequency response of the inverse filter has the form

$$\frac{1}{\frac{1}{2} + \frac{1}{4}e^{-j2\pi\alpha}} = 2\left(1 - \frac{1}{2}e^{-j2\pi\alpha} + \frac{1}{4}e^{-j4\pi\alpha} - \frac{1}{8}e^{-j6\pi\alpha} + \cdots\right). \tag{6.10}$$

The coefficients of the series expansion (6.10) are equal to the sample values of the filter pulse response. It is an IIR filter, with infinitely many coefficients. We should avoid the use of IIR inverse (synthesis) filters in multimedia compression systems since a quantization error introduced by quantization of the transform coefficients can be amplified by feedback. Thus, in this case we can find the inverse transform matrix T^{-1} but we cannot implement it as a FIR filtering.

6.3   "Moving difference" filtering as a linear transform

Now we consider a highpass filter which is described by the following equation

$$y(n) = \frac{1}{2}x(n) - \frac{1}{2}x(n-1). \tag{6.11}$$

This FIR filter is called a *moving difference* filter. The length of its pulse response is 2. The filter coefficients are $h(0) = \frac{1}{2}$ and $h(1) = -\frac{1}{2}$. The corresponding $N \times N$ transform matrix has the form

$$T = \begin{pmatrix} \frac{1}{2} & 0 & 0 & \cdots & 0 & -\frac{1}{2} \\ -\frac{1}{2} & \frac{1}{2} & 0 & 0 & \cdots & 0 \\ 0 & -\frac{1}{2} & \frac{1}{2} & 0 & \cdots & 0 \\ \cdots & \cdots & \cdots & \cdots & \cdots & \cdots \\ 0 & 0 & \cdots & 0 & -\frac{1}{2} & \frac{1}{2} \end{pmatrix}. \tag{6.12}$$

The frequency response of this highpass filter is

$$H(e^{j\omega T_s}) = \sum_n h(n)e^{-j\omega n T_s} = \frac{1}{2} - \frac{1}{2}e^{-j\omega T_s}.$$

As before, we take out a factor $e^{-j\omega T_s/2}$ and obtain the frequency response in the form

$$H(e^{j\omega T_s}) = \sin\left(\frac{\omega T_s}{2}\right) j e^{-j\omega T_s/2}. \tag{6.13}$$

Thus, the amplitude is

$$H(\omega) = \left|\sin\left(\frac{\omega T_s}{2}\right)\right|, \ H(\alpha) = |\sin(\pi\alpha)|.$$

The phase is

$$\varphi(\omega) = \begin{cases} \frac{\pi}{2} - \frac{\omega T_s}{2} & \text{for } 0 < \omega < \pi/T_s \\ -\frac{\pi}{2} - \frac{\omega T_s}{2} & \text{for } -\pi/T_s < \omega < 0 \end{cases}$$

$$\varphi(\alpha) = \begin{cases} \frac{\pi}{2} - \pi\alpha & \text{for } 0 < \alpha < \frac{1}{2} \\ -\frac{\pi}{2} - \pi\alpha & \text{for } -\frac{1}{2} < \alpha < 0. \end{cases}$$

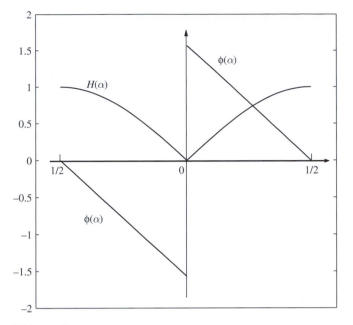

Figure 6.2   Highpass filter

The plots of $H(\alpha)$ and $\varphi(\alpha)$ are shown in Fig. 6.2. It is easy to see that the filter has zero response to the sequence $(\ldots, 1, 1, 1, 1, 1, \ldots)$ and unit response to the sequence $(\ldots, 1, -1, 1, -1, 1, \ldots)$. In other words, it preserves the highest discrete-time frequency component and it eliminates the DC term of the input signal.

The phase of the filter has a discontinuity. It jumps from $-\pi/2$ to $\pi/2$ at $\alpha = 0$. At the other points the graph is linear and we do not pay attention to this discontinuity and say that the filter is still a linear phase filter. The moving difference filter is also not invertible since its amplitude $H(\alpha) = 0$ at $\alpha = 0$.

6.4   The Haar filter bank

In order to distinguish between the frequency responses of the lowpass and highpass filters let us denote as H_0 the frequency response (6.8) and as H_1 the frequency response (6.13). Let T_0 and T_1 be the corresponding transform matrices (6.6) and (6.12).

It follows from the above consideration that separately, the moving average lowpass and the moving difference highpass filters are not invertible. Moreover, even if the inverse filter exists, it very rarely turns out that this filter is a FIR filter since $1/H(e^{j2\pi\alpha})$ is not a polynomial. In order to construct an invertible discrete transform which can be implemented by using FIR filters, we need to go to *filter banks*.

Considered together the above filters separate the input sequence x into frequency bands in such a way that $T_1 x$ is the complement of $T_0 x$. It is said that these filters form a filter bank. It can be shown that there exists an inverse filter bank or, in other words, the transform which splits the input signal into low-frequency and high-frequency components is invertible. Moreover, the inverse filter bank contains FIR filters; that is, the inverse transform can be implemented via FIR filtering.

However, there is one more difficulty related to the transform based on filter banks. It is easy to see that this transform doubles the length of the output sequence compared to the input sequence since both lowpass and highpass filtered sequences have a length equal to the length of the input sequence. The solution to this problem is to *downsample* (or *decimate*) the output sequences.

Let the outputs of the filters be $y_0 = T_0 x$; $y_1 = T_1 x$; then decimation means removing all odd-numbered components of y_0 and y_1. Downsampling is represented by the symbol $(\downarrow 2)$:

$$(\downarrow 2) \begin{pmatrix} y(0) \\ y(1) \\ y(2) \\ y(3) \\ \cdots \end{pmatrix} = \begin{pmatrix} y(0) \\ y(2) \\ y(4) \\ y(6) \\ \cdots \end{pmatrix}.$$

This operation is not invertible. The odd-numbered components of both (low-frequency and high-frequency) components y_0 and y_1 are lost. It looks amazing but we can recover the input vector x from the even-numbered components of y_0 and y_1. An intuitive explanation is the following. Assume that the filter bank contains the ideal lowpass and highpass filters. Amplitudes of their frequency responses are rectangle functions which divide the original frequency band into two subbands of the same width. According to the sampling theorem we can sample the corresponding lowpass and highpass filtered signals with half the sampling frequency without introducing any distortions.

In order to take into account downsampling, we multiply the left $y(2n)$ by $\sqrt{2}$ and replace $H_0(\alpha)$ by $\sqrt{2}H_0(\alpha)$, and $H_1(\alpha)$ by $\sqrt{2}H_1(\alpha)$. The corresponding transform matrices T_0 and T_1 are replaced by the matrices T_L and T_H, respectively. It is easy to see that both T_L and T_H are matrices with a double shift. They consist of 1×2 blocks and no longer have constant diagonals. All entries of the matrices are multiplied by the normalizing factor $\sqrt{2}$.

The two steps, filtering and decimation, can be done with each of these new matrices of size $N/2 \times N$

$$T_L = \frac{1}{\sqrt{2}} \begin{pmatrix} 1 & 0 & \cdots & 0 & 0 & 0 & 1 \\ 0 & 1 & 1 & 0 & 0 & \cdots & 0 \\ 0 & 0 & 0 & 1 & 1 & 0 & \cdots \\ \cdots & \cdots & \cdots & \cdots & \cdots & \cdots & \cdots \end{pmatrix}$$

and

$$T_H = \frac{1}{\sqrt{2}} \begin{pmatrix} 1 & 0 & \cdots & 0 & 0 & 0 & -1 \\ 0 & -1 & 1 & 0 & 0 & \cdots & 0 \\ 0 & 0 & 0 & -1 & 1 & 0 & \cdots \\ \cdots & \cdots & \cdots & \cdots & \cdots & \cdots & \cdots \end{pmatrix}.$$

The rectangular matrices T_L and T_H form a square transform matrix of size $N \times N$ which represents the complete *analysis filter bank*

$$\begin{bmatrix} T_L \\ T_H \end{bmatrix} = \frac{1}{\sqrt{2}} \begin{pmatrix} 1 & 0 & 0 & \ldots & 0 & 1 \\ 0 & 1 & 1 & 0 & \ldots & 0 \\ \ldots & \ldots & \ldots & \ldots & \ldots & \ldots \\ 1 & 0 & 0 & \ldots & 0 & -1 \\ 0 & -1 & 1 & 0 & \ldots & 0 \\ \ldots & \ldots & \ldots & \ldots & \ldots & \end{pmatrix}.$$

The combined square matrix is invertible. Moreover, the inverse matrix is equal to the transposed matrix of the forward transform (analysis filter bank), that is,

$$\begin{bmatrix} T_L \\ T_H \end{bmatrix}^{-1} = \begin{bmatrix} T_L^T & T_H^T \end{bmatrix} = \frac{1}{\sqrt{2}} \begin{pmatrix} 1 & 0 & \vdots & 1 & 0 & \vdots \\ 0 & 1 & \vdots & 0 & -1 & \vdots \\ 0 & 1 & \vdots & \vdots & 1 & \vdots \\ \vdots & 0 & \vdots & 0 & 0 & \vdots \\ 0 & \vdots & \vdots & 0 & \vdots & \vdots \\ 1 & 0 & \vdots & -1 & 0 & \vdots \end{pmatrix}.$$

It is easy to check that

$$\begin{bmatrix} T_L^T & T_H^T \end{bmatrix} \begin{bmatrix} T_L \\ T_H \end{bmatrix} = T_L^T T_L + T_H^T T_H = I.$$

The matrix

$$\begin{bmatrix} T_L^T & T_H^T \end{bmatrix} \tag{6.14}$$

represents the inverse transform which corresponds to the *synthesis filter bank*. Since the inverse matrix is equal to the transposed matrix of the forward transform we can say that the obtained transform is orthogonal.

We are going to implement the inverse transform via a filter bank. For this purpose the inverse transform is organized to have two steps: *upsampling* and *filtering*.

First, we obtain full-length vectors from the decimated vectors. It is performed by inserting zero odd-numbered components into half-length transformed vectors. The operation is denoted by ($\uparrow$ 2) and called *upsampling*:

$$(\uparrow 2) \begin{pmatrix} v(0) \\ v(1) \\ v(2) \\ \ldots \\ \ldots \end{pmatrix} = \begin{pmatrix} v(0) \\ 0 \\ v(1) \\ 0 \\ v(2) \\ 0 \\ \ldots \end{pmatrix}.$$

In such a way we obtain the extended vectors $\tilde{y}_0$ and $\tilde{y}_1$. The second step of the inverse (synthesis) transform is filtering. The lowpass filtering is equivalent to multiplying the extended vector $\tilde{y}_0$ by the matrix:

$$S_0 = \frac{1}{\sqrt{2}} \begin{pmatrix} 1 & 1 & 0 & \ldots & 0 & 0 \\ 0 & 1 & 1 & 0 & \ldots & 0 \\ 0 & 0 & 1 & 1 & 0 & \ldots \\ \ldots & \ldots & \ldots & \ldots & \ldots & \ldots \\ 1 & 0 & 0 & \ldots & 0 & 1 \end{pmatrix}. \tag{6.15}$$

The highpass filtering is equivalent to multiplying the extended vector $\tilde{y}_1$ by the matrix:

$$S_1 = \frac{1}{\sqrt{2}} \begin{pmatrix} 1 & -1 & 0 & \ldots & 0 & 0 \\ 0 & 1 & -1 & 0 & \ldots & 0 \\ 0 & 0 & 1 & -1 & 0 & \ldots \\ \ldots & \ldots & \ldots & \ldots & \ldots & \ldots \\ -1 & 0 & 0 & \ldots & 0 & 1 \end{pmatrix}. \tag{6.16}$$

It was shown that linear filtering is a linear transform described by matrix (6.4). However, in order to match forward and inverse filterings, we have to start filtering an extended periodical sequence with the sample $y(0)$ and stop after $N + l - 1$ steps. Then we take the samples $y(l - 1), y(l - 1), y(l) \ldots, y(N + l - 2)$ of the output sequence $x(n)$. Such a filtering can be described by the matrix

$$T = \begin{pmatrix} h(l-1) & h(l-2) & \ldots & h(0) & 0 & \ldots & 0 \\ 0 & h(l-1) & h(l-2) & \ldots & h(0) & 0 & \ldots \\ \ldots & \ldots & \ldots & \ldots & \ldots & \ldots & \ldots \\ h(l-2) & \ldots & h(0) & 0 & \ldots & 0 & h(l-1) \end{pmatrix}. \tag{6.17}$$

Comparing matrices (6.15) and (6.16) with the matrix (6.17), we find that the synthesis lowpass and highpass filters are determined as follows:

$$y(n) = \frac{1}{\sqrt{2}} (x(n) + x(n-1)),$$

$$y(n) = \frac{1}{\sqrt{2}} (x(n-1) - x(n)).$$

The sum $S_0 \tilde{y}_0 + S_1 \tilde{y}_1$ represents the delayed input signal $\tilde{x}(n) = x(n - l + 1) = x(n - 1)$. The above filter bank is called the *Haar filter bank*.

Notice that when the analysis and synthesis filter banks are not orthogonal, then pulse responses of the lowpass and highpass filter can be of different lengths. In this case matrices (6.4) and (6.17) have another form.

Example 6.1 Let

$$x(n) = x(0), x(1), x(2), x(3), x(4), x(5)$$

be the input sequence of the Haar filter bank. The length of the input sequence is $N = 6$. The length of the pulse response for both filters is $l = 2$.

The periodically extended input sequence is

$$x_p(n) = \ldots, x(5), x(0), x(1), x(2), x(3), x(4), x(5), x(0), \ldots$$

The periodical infinite low-frequency output sequence is

$$\ldots, x(4) + x(5), x(0) + x(5), x(1) + x(0), x(1) + x(2), \ldots, x(4) + x(5),$$

$$x(0) + x(5), \ldots$$

The low-frequency output sequence multiplied by the window is

$$y_0(n) = x(0) + x(5), x(0) + x(1), x(1) + x(2), x(2) + x(3), x(3) + x(4),$$

$$x(4) + x(5).$$

The periodical infinite high-frequency output sequence is

$$\ldots, x(5) - x(4), x(0) - x(5), x(1) - x(0), x(2) - x(1), \ldots, x(5) - x(4),$$

$$x(0) - x(5), \ldots$$

The high-frequency output sequence multiplied by the window is

$$y_1(n) = x(0) - x(5), x(1) - x(0), x(2) - x(1), x(3) - x(2), x(4) - x(3),$$
$$x(5) - x(4).$$

The decimated low-frequency part is

$$v_0(n) = x(0) + x(5), x(1) + x(2), x(3) + x(4).$$

The decimated high-frequency part is

$$v_1(n) = x(0) - x(5), x(2) - x(1), x(4) - x(3).$$

The upsampled low-frequency part is

$$\tilde{y}_0(n) = x(0) + x(5), 0, x(1) + x(2), 0, x(3) + x(4), 0.$$

The upsampled high-frequency part is

$$\tilde{y}_1(n) = x(0) - x(5), 0, x(2) - x(1), 0, x(4) - x(3), 0.$$

The periodically extended low-frequency part is

$$\ldots, 0, x(0) + x(5), 0, x(1) + x(2), 0, x(3) + x(4), 0, x(0) + x(5), \ldots \quad (6.18)$$

The periodically extended high-frequency part is

$$\ldots, 0, x(0) - x(5), 0, x(2) - x(1), 0, x(4) - x(3), 0, x(0) - x(5), \ldots \qquad (6.19)$$

The periodical infinite sequence (6.18) filtered by the lowpass reconstruction filter has the form

$$w_0(n) = \ldots, x(0) + x(5), x(0) + x(5), x(1) + x(2), x(1) + x(2), x(3) + x(4),$$
$$x(3) + x(4), x(0) + x(5), \ldots$$

The periodical infinite sequence (6.19) filtered by the highpass reconstruction filter has the form

$$w_1(n) = \ldots, x(5) - x(0), x(0) - x(5), x(1) - x(2), x(2) - x(1), x(3) - x(4),$$
$$x(4) - x(3), x(5) - x(0).$$

Summing up $w_0(n)$ and $w_1(n)$ after normalization, we obtain

$$\ldots, x(5), x(0), x(1), x(2), x(3), x(4), x(5), \ldots \qquad (6.20)$$

By windowing the sequence (6.20), we obtain

$$\tilde{x}(n) = x(n - 1).$$

The corresponding forward and inverse transform matrices have the form

$$[T_\mathrm{L} T_\mathrm{H}] = \frac{1}{\sqrt{2}} \begin{pmatrix} 1 & 0 & 0 & 0 & 0 & 1 \\ 0 & 1 & 1 & 0 & 0 & 0 \\ 0 & 0 & 0 & 1 & 1 & 0 \\ 1 & 0 & 0 & 0 & 0 & -1 \\ 0 & -1 & 1 & 0 & 0 & 0 \\ 0 & 0 & 0 & -1 & 1 & 0 \end{pmatrix} \qquad (6.21)$$

$$[T_\mathrm{L}^\mathrm{T} T_\mathrm{H}^\mathrm{T}] = \frac{1}{\sqrt{2}} \begin{pmatrix} 1 & 0 & 0 & 1 & 0 & 0 \\ 0 & 1 & 0 & 0 & -1 & 0 \\ 0 & 1 & 0 & 0 & 1 & 0 \\ 0 & 0 & 1 & 0 & 0 & -1 \\ 0 & 0 & 1 & 0 & 0 & 1 \\ 1 & 0 & 0 & -1 & 0 & 0 \end{pmatrix}. \qquad (6.22)$$

Notice that in practice it is not necessary to consider an infinite periodically extended input sequence. It is enough to place $l - 1$ last samples into the beginning of the input sequence (or $l - 1$ first samples to the end of the input sequence). This method is called *cyclic extension*.

6.5   Wavelet transform and wavelet filter bank

As already mentioned, the DFT and DCT have common disadvantages that both transforms provide a good frequency resolution (good localization of energy) only for long
harmonic functions. On the other hand, multimedia signals such as images can contain
contours or small details; similarly, audio signals can contain short sounds or so-called
sound attacks and we need a transform which would represent these short time functions
efficiently. Typically, we need a good frequency resolution for low-frequency signals
and a good time resolution for high-frequency signals. Wavelet transforms have advantages compared to Fourier transforms when representing finite, nonperiodic, and/or
nonstationary signals. The output of the wavelet transform is a set of time-frequency
representations of the input signal with different resolutions.

The wavelet filter banks have special properties. The most important feature of these
filter banks is their hierarchical structure. In fact, filtering an input vector by a wavelet
filter bank represents an implementation of the *Discrete Wavelet Transform* (DWT). In
other words, wavelet filtering implements a decomposition of the input vector over a
set of basis functions called *wavelets*. The recursive nature of wavelets which will be
explained below leads to a tree structure of the corresponding filter bank.

We will start by describing the Continuous Wavelet Transform (CWT).

For any admissible continuous function $f(t)$, CWT is defined as

$$b(s, \tau) = \int_{-\infty}^{\infty} f(t) \psi_{s,\tau}(t) \, dt$$

where

$$\psi_{s,\tau}(t) = \frac{1}{\sqrt{s}} \psi\left(\frac{t - \tau}{s}\right), \, s, \tau \in \mathcal{R}, s \neq 0$$

are basis functions called *wavelets*, s is the scale factor, and τ is the translation factor.
It is easy to see that the basis functions $\psi_{s,\tau}(t)$ are generated by scaling and translation
from a single basis wavelet $\psi(t)$. It is called the *mother wavelet*.

The term "admissible function" means a function from the space $\mathcal{L}^2$ (space with the
$\mathcal{L}^2$ norm, that is,

$$\|f(t)\| = \left(\int_{-\infty}^{\infty} |f(t)|^2 \, dt\right)^{1/2}$$

is finite). The inverse wavelet transform is determined by the formula

$$f(t) = \int_{-\infty}^{\infty} \int_{-\infty}^{\infty} b(s, \tau) \psi_{s,\tau}(t) \, d\tau \, ds.$$

In the wavelet theory, wavelet functions themselves are not specified; only their properties (for details, see, for example, (Strang and Nguyen 1996)) are determined. These
properties guarantee the existence of an invertible transform. The term wavelet means
a *small wave*. One of the required properties is that the average value of the wavelet in
the time domain has to be zero; that is,

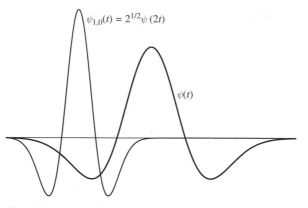

Figure 6.3   Classical wavelet: "Mexican hat" and its scaled version

$$\int_{-\infty}^{\infty} \psi(t)\, dt = 0.$$

It means that any wavelet function has to contain oscillations; that is, it resembles a wave. A classical example of a wavelet is the *Mexican hat* function shown in Fig. 6.3,

$$\psi(t) = (1 - 2t^2) \exp(-t^2). \tag{6.23}$$

As we will see later, in order to represent any function from $\mathcal{L}^2$ we do not need a continuous basis where s and τ vary continuously. The first step towards more practical transforms is to use, instead of the continuous wavelet basis, the so-called *discrete wavelets*. They are not continuously scalable and translatable but can only be scaled and translated in discrete steps

$$\psi_{j,k}(t) = \frac{1}{\sqrt{s_0^{-j}}} \psi\left(\frac{t - k\tau_0 s_0^{-j}}{s_0^{-j}}\right), \; j,k \in Z.$$

Usually, the value τ_0 is equal to 1 and the value s_0 is equal to 2; that is,

$$\psi_{j,k}(t) = \sqrt{2^j}\, \psi(2^j t - k).$$

It is said that in this case we deal with *dyadic sampling* of the time and frequency axes. Normally, wavelets start at time $t = 0$ and end at time $t = T$. The shifted wavelets $\psi_{0,k}(t)$ start at time $t = k$ and end at time $t = k + T$. The rescaled wavelets $\psi_{j,0}(t)$ start at time $t = 0$ and end at time $t = T/2^j$. Their graphs are shortened (if $j > 0$) by the factor 2^j, whereas the graphs of $\psi_{0,k}(t)$ are translated (shifted to the right) by k. In Fig. 6.3, we also show the scaled Mexican hat function

$$\psi_{1,0}(t) = \sqrt{2}\psi(2t)$$

where $\psi(t)$ is determined by (6.23). We see that the wavelet basis contains rather short time functions that allow us to represent short input time functions efficiently.

A continuous time function $f(t)$ can be decomposed using discrete wavelets as follows:

$$f(t) = \sum_{j,k} b_{j,k} \psi_{j,k}(t), \qquad (6.24)$$

where

$$b_{j,k} = \int_{-\infty}^{\infty} f(t) \psi_{j,k}(t) \, dt. \qquad (6.25)$$

Notice that discrete wavelets are continuous time basis functions. Only the translation and the scale steps are discrete. Often the series of wavelet coefficients $b_{j,k}$ in (6.24) are referred to as the *wavelet series decomposition*.

The equality (6.24) can be rewritten as

$$f(t) = \sum_{j} w_j(t) \qquad (6.26)$$

where $w_j(t) = \sum_k b_{j,k} \psi_{j,k}(t)$ is an approximation of the function $f(t)$ in the subspace $W_j \subset \mathcal{L}^2$ generated by the basis functions $\psi_{j,k}(t)$ for a given j. It is said that (6.24) corresponds to the following equality for functional spaces

$$\mathcal{L}^2 = \sum_{j} \dot{W}_j$$

where $\sum$ denotes the direct sum of subspaces. (The basis of the direct sum of subspaces is a union of bases of these subspaces. If the subspaces are orthogonal, then the dimension of the direct sum is equal to the sum of the dimensions of the subspaces; otherwise it is less than the sum of the dimensions.)

From the implementation point of view, both transforms (continuous wavelet transform and wavelet series decomposition) presented above are impractical since the wavelet basis functions, no matter if they are continuous or discrete, have a rather complicated analytical description or can even have no analytical description and can be implemented only numerically.

Assume that our continuous time function is discretized and we deal with a sequence of samples $f(n)$, $n = 1, 2, \ldots, N$. Consider the DWT which can be applied to $f(n)$. A standard approach to go from a continuous to a discrete transform is to discretize the corresponding basis functions. In such a way we can obtain, for example, the DFT from the continuous Fourier transform. In the case of wavelet transforms, instead of discretizing basis functions we use another approach to arrive at the DWT. This approach is called *multiresolution analysis* (Mallat 1999). The main idea behind this approach is to compute the wavelet coefficients $b_{j,k}$ (6.25) recurrently; that is, to express the wavelet coefficients at the jth resolution level via the wavelet coefficients at the resolution level $j - 1$. Moreover, it can be shown that formulas connecting wavelet coefficients for different resolution levels can be implemented as iterated filtering by a filter bank. This implementation does not require specification of the wavelet basis functions at all.

An important observation underlying the multiresolution approach is that the mother wavelet has a band-pass-like spectrum. A wavelet $\psi_{j,k}(t)$ has a spectrum which is stretched (if $j > 0$) by a factor of $s_0^j = 2^j$ with respect to the spectrum of the mother wavelet and its central frequency is shifted up by a factor of $s_0^j = 2^j$. Thus, decomposition of an admissible function over wavelet basis functions can be considered as a filtering by a band-pass filter bank where each next filter has twice as wide a frequency band as the previous one.

Let us introduce the so-called *scaling function*, which is a signal $\varphi(t)$ with a low-frequency spectrum. It can be decomposed using the wavelet basis

$$\varphi(t) = \sum_{j,k} b_{j,k} \psi_{j,k}(t)$$

where $b_{j,k} \in \mathcal{R}$ are wavelet transform coefficients. The scaling function is also called the *father wavelet* since, similar to the mother wavelet $\psi(t)$, it can be used to generate a set of scaled and translated scaling functions $\varphi_{j,k}(t)$

$$\varphi_{j,k}(t) = \frac{1}{\sqrt{s_0^{-j}}} \varphi\left(\frac{t - k\tau_0 s_0^{-j}}{s_0^{-j}}\right).$$

The scaling functions $\sqrt{2^j}\varphi(2^j t - k)$ with a given j form the basis of the subspace $V_j \subset \mathcal{L}^2$. The chain of subspaces V_j satisfies the following conditions

$$\{0\} \subset \ldots \subset V_j \subset V_{j+1} \ldots \subset \mathcal{L}^2 \tag{6.27}$$

$$V_{j+1} = W_j \dotplus V_j \tag{6.28}$$

where (6.28) means that the wavelet subspace at level j is a complementary subspace with respect to the scaling subspace V_j.

Decomposition of $f(t)$ using a basis containing a scaling function represented via wavelets up to a given $j = J$ and dilated wavelets with $j \geq J + 1$ can also be interpreted as filtering by a filter bank. One of the filters corresponding to the scaling function is a lowpass filter and the other wavelet filters are band-pass filters.

Denote by $v_j(t) = \sum_k a_{j,k} \varphi_{j,k}(t)$ an approximation of $f(t)$ in V_j. Then, taking into account (6.28), we can rewrite (6.26) as

$$f(t) = v_{J+1}(t) + \sum_{j=J+1} w_j(t).$$

Now consider a scaling function which can be represented via wavelets up to a certain scale. It follows from (6.27) and (6.28) that if we add a wavelet spectrum to the scaling function spectrum, we will obtain the spectrum of a new scaling function which is twice as wide as the spectrum of the initial scaling function.

By using the recurrent nature of subspaces V_j and approximations $v_j(t)$, it is possible to express decomposition coefficients for an approximation $v_{j+1}(t)$ of the input function with resolution level $j + 1$ via decomposition coefficients for an approximation $v_j(t)$ of the input sequence with a previous resolution level j. It can be shown that computing decomposition coefficients of the given level via decomposition coefficients of the previous level is equivalent to filtering a sequence of input decomposition coefficients by FIR filters. A lowpass filter corresponds to the decomposition coefficients in subspace V_j and a highpass filter corresponds to the decomposition coefficients in the complementary subspace W_j. Notice that with respect to the subspace V_{j+1}, the wavelet band-pass filter acts as a highpass filter. This approach requires a specification of neither a scaling function nor a wavelet. Only the pulse responses of the filters have to be specified.

Since $V_0 \subset V_1$, the scaling function $\varphi(t) \in V_0$ can be represented as a linear combination of basis functions from V_1,

$$\varphi(t) = \sqrt{2} \sum_k h_{0,k} \varphi_{1,k}(t) \tag{6.29}$$

where $h_{0,k}$ are weighting coefficients. Since $W_0 \subset V_1$, the wavelet $\psi(t) \in W_0$ can be represented as

$$\psi(t) = \sqrt{2} \sum_k h_{1,k} \varphi_{1,k}(t) \tag{6.30}$$

where $h_{1,k}$ are weighting coefficients. Equations (6.29) and (6.30) are called the *dilation* and the *wavelet* equations, respectively.

By a straightforward substitution of indices in (6.29) and (6.30) and taking into account that the basis functions of V_{j+1} are half the width of the basis functions of V_j, we obtain that the scaling function $\varphi_{j,l}(t)$ from V_j can be represented as a linear combination of translated scaling functions from V_{j+1},

$$\varphi_{j,l}(t) = \sqrt{2} \sum_k h_{0,j+1,k} \varphi_{j+1,k}(t) = \sqrt{2} \sum_k h_{0,k-2l} \varphi_{j+1,k}(t).$$

Analogously, the wavelet function $\psi_{j,l}(t)$ from W_j can also be expressed via translated scaling functions from V_{j+1},

$$\psi_{j,l}(t) = \sqrt{2} \sum_k h_{1,j+1,k} \varphi_{j+1,k}(t) = \sqrt{2} \sum_k h_{1,k-2l} \varphi_{j+1,k}(t).$$

An approximation $v_{J+1}(t) \in V_{J+1}$ can be expressed in two ways:

$$v_{J+1}(t) = \sum_k a_{J,k} \varphi_{J,k}(t) + \sum_k b_{J,k} \psi_{J,k}(t) \tag{6.31}$$

and

$$v_{J+1}(t) = \sum_k a_{J+1,k} \varphi_{J+1,k}(t) \tag{6.32}$$

where

$$a_{J,k} = \int_{-\infty}^{\infty} v_{J+1}(t) \varphi_{J,k}(t)\, dt$$

$$b_{J,k} = \int_{-\infty}^{\infty} v_{J+1}(t)\psi_{J,k}(t)\,dt.$$

Equating (6.31) and (6.32) and taking into account that the scaling functions $\varphi_{j,k}(t)$ and the wavelets $\psi_{j,k}(t)$ are orthonormal, we obtain

$$a_{j,k} = \sum_n h_0(n - 2k)a_{j+1,k} \qquad (6.33)$$

and

$$b_{j,k} = \sum_n h_1(n - 2k)a_{j+1,k} \qquad (6.34)$$

where $h_0(n - 2k) = h_{0,n-2k}$ and $h_1(n - 2k) = h_{1,n-2k}$. Equations (6.33) and (6.34) describe an iterative procedure for finding the coefficients of the DWT of $f(k)$, $k = 1, 2, \ldots, N$. We assume that the samples of the input sequence $f(k) = a_{j,k}$ for the largest j. At each step of the decomposition we compute a new set of coefficients $a_{j,k}$ which can be considered as time-frequency representations of the input sequence with the resolution level j. The coefficients $h_0(n)$ and $h_1(n)$ can be interpreted as samples of the pulse response of the lowpass and highpass filter, respectively. A step size of 2 in the variable k is implemented as downsampling of the filtered low-frequency and high-frequency sequences.

6.6   Hierarchical wavelet filtering as a linear transform. Properties of wavelet filtering

Now we consider the iterative procedure described by (6.33) and (6.34) in more detail. The input signal is decomposed using two filters into low-frequency and high-frequency parts. Then, each component of the input signal is decimated; that is, the only even-numbered samples are kept. The downsampled high-frequency part represents a final output because it is not transformed again. Since this part of the signal typically contains a rather insignificant part of the signal energy, it can be encoded by using a small number of bits. The decimated low-frequency component usually contains the main part of the signal energy and it is filtered again by the same pair of filters. The decimated high-frequency part of the low-frequency signal component is not transformed again but the decimated low-frequency part of the low-frequency signal component can be filtered again and so on.

Thus, the DWT is a hierarchical decomposition of the input sequence into the so-called *reference* (low-frequency) subsequences with diminishing resolutions and related with them the so-called *detail* (high-frequency) subsequences. At each level of decomposition, the DWT is invertible; that is, the reference signal of this level together with the corresponding detail signal provides perfect reconstruction of the reference signal of the next level (with higher resolution). Below, we use notations commonly used in the image compression area. Namely, we denote as $r_j(n)$ the reference signal of the jth level of the decomposition and as $d_j(n)$ the detail signal of the jth level.

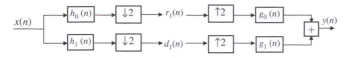

Figure 6.4   One level of wavelet decomposition followed by reconstruction

Figure 6.4 illustrates one level of wavelet decomposition followed by a reconstruction. The input sequence $x(n)$ is filtered by a lowpass filter with the pulse response $h_0(n)$ and by a highpass filter with the pulse response $h_1(n)$. The downsampling step is symbolized by ($\downarrow 2$). The sequence $r_1(n)$ is the reference signal (decimated result of lowpass filtering), and $d_1(n)$ is the detail signal (decimated result of highpass filtering). It is evident that this scheme transforms one sequence of length N into two subsequences of length $N/2$ each.

In the theory of wavelet filter banks, such pairs of filters $h_0(n)$ and $h_1(n)$ are found that there exist pairs of the inverse FIR filters $g_0(n)$ and $g_1(n)$ providing the perfect reconstruction of the input signal. To reconstruct the input signal from signals $r_1(n)$ and $d_1(n)$, these signals are first upsampled with a factor of 2. In Fig. 6.4 upsampling is symbolized by ($\uparrow 2$). Then, the upsampled low-frequency and high-frequency components are filtered by the inverse lowpass filter with the pulse response $g_0(n)$ and the inverse highpass filter with the pulse response $g_1(n)$, respectively. The sum of the results of filtering is the output signal $y(n)$. The DWT (wavelet filtering) provides a perfect reconstruction of the input signal; that is, the output signal is determined as

$$y(n) = Ax(n - \alpha)$$

where A is the gain factor and α is the delay.

In the case of multilevel decomposition the reference signal $r_1(n)$ represents the input signal of the next decomposition level. Filtering is performed iteratively as shown in Fig. 6.5. At the Lth level of decomposition we obtain the reference signal $r_L(n)$ with resolution 2^L times scaled down compared to the resolution of the input signal and the detail signals $d_L(n), d_{L-1}(n), \ldots, d_1(n)$ with resolution 2^j, $j = L, L - 1, \ldots, 1$, times scaled down compared to the input signal, respectively. Each detail signal $d_i(n)$ contains such information that together with the reference signal $r_i(n)$ it allows recovering of $r_{i-1}(n)$ which represents the reference signal of the next level. At the Lth level of the decomposition, the total length of the reference and detail subsequences is

$$2^{-L}N + 2^{-L}N + 2^{-(L-1)}N + 2^{-(L-2)}N + \cdots + 2^{-1}N$$

$$= N \left(\sum_{i=1}^{L} 2^{-i} + 2^{-L} \right) = N \left(2^{-1} \frac{1 - 2^{-L}}{1 - 2^{-1}} + 2^{-L} \right) = N.$$

Notice that besides the hierarchical structure, wavelet filter banks have a second very important property, namely, that synthesis wavelet filter banks always consist of FIR linear phase filters, which is very convenient from the implementation point of view.

The above-mentioned remarkable properties of wavelet filter banks follow from the dilation and wavelet equations (6.29) and (6.30) which connect pulse responses of

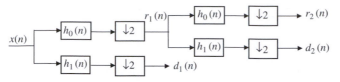

Figure 6.5   Multiresolution wavelet decomposition

wavelet filters $h_0(n)$ and $h_1(n)$ with scaling function $\varphi(t)$ and wavelet $\psi(t)$. Similarly, for synthesis filter banks with pulse responses $g_0(n)$ of the lowpass filter and $g_1(n)$ of the highpass filter, we have

$$\varphi_s(t) = \sum_k g_0(k)\varphi_s(2t - k) \tag{6.35}$$

$$\psi_s(t) = \sum_k g_1(k)\varphi_s(2t - k) \tag{6.36}$$

where $\varphi_s(t)$, $\psi_s(t)$ denote the synthesis scaling function and the synthesis wavelet. By choosing different wavelet and scaling functions, one can govern the performances of analysis and synthesis filter banks.

We can summarize that the DWT decomposes an input signal into different scales of resolution, rather than different frequencies. When using a multiresolution approach, the frequency band of the input signal is divided into subbands with bandwidths corresponding to the decomposition level, instead of uniform subbands as the DFT does. At each level of the wavelet decomposition the time steps are reduced by a factor of 2 and the frequency steps are doubled.

Now we return to the Haar filter bank. It can be shown that the Haar filter bank is an example of a wavelet filter bank. For a lowpass filter with $h_0(0) = \frac{1}{2}$ and $h_0(1) = \frac{1}{2}$, we obtain the dilation equation:

$$\varphi(t) = \varphi(2t) + \varphi(2t - 1). \tag{6.37}$$

It is easy to see that the solution is the *box function*:

$$\varphi(t) = \begin{cases} 1, & \text{for } 0 \leq t \leq 1 \\ 0, & \text{otherwise} \end{cases} \tag{6.38}$$

which is equal to the sum of two half-size boxes $\varphi(2t)$ and $\varphi(2t - 1)$. The functions $\varphi(t)$, $\varphi(2t)$, and $\varphi(2t - 1)$ are shown in Fig. 6.6.

The coefficients of the wavelet equation for our example are $h_1(0) = \frac{1}{2}$ and $h_1(1) = -\frac{1}{2}$. Taking into account that $\varphi(2t)$ and $\varphi(2t - 1)$ are half-boxes, we obtain that the wavelet is a difference of half-boxes:

$$\psi(t) = \varphi(2t) - \varphi(2t - 1) \tag{6.39}$$

that is, $\psi(t) = 1$ for $0 \leq t \leq \frac{1}{2}$ and $\psi(t) = -1$ for $\frac{1}{2} \leq t < 1$. This is the *Haar wavelet*. Its graph is shown in Fig. 6.7 together with the graphs of $\psi(2t)$ and $\psi(2t - 1)$. We mentioned before that a wavelet is a small wave. In Haar's case it is a square wave. Notice that Alfred Haar was writing in 1910 about this function without calling it "wavelet." The Haar wavelet is zero outside a bounded interval [0,1]. It is said that it has a "compact

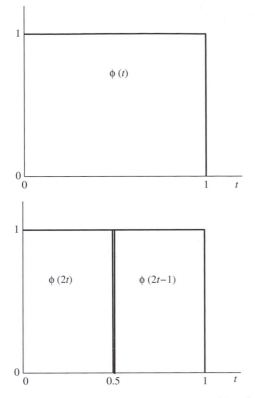

Figure 6.6   Box scaling function. Scaled and translated box functions

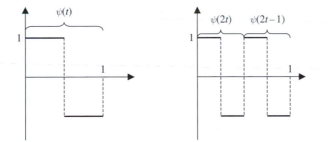

Figure 6.7   The Haar wavelet $\psi(t)$. The scaled and translated wavelets $\psi(2t)$ and $\psi(2t-1)$

support." Due to this property the corresponding filters are FIR filters. Haar's wavelet basis containing all functions $\psi(2^j - k)$, $j, k \in Z$, is an orthonormal basis and the corresponding transform is orthonormal.

Let us apply the Haar filter bank recurrently as shown in Fig. 6.8. For our example the highpass filter computes the differences of the input samples. The downsampling step keeps the even-numbered differences $(x(2k) - x(2k - 1))/\sqrt{2}$ which are the final outputs b_{jk} because they are not transformed again. The factor $1/\sqrt{2}$ is included for normalization. The lowpass filter computes the averages of the input samples. The downsampling keeps the even samples $(x(2k - 1) + x(2k))/\sqrt{2}$. These averages a_{jk} are not the final outputs, because they will be filtered again. It is easy to see that

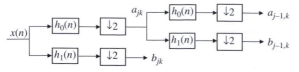

Figure 6.8   The logarithmic tree structure of the wavelet filter bank

the averages and differences at all levels of the decomposition satisfy the following recurrent equations

$$a_{jk} = \frac{1}{\sqrt{2}} \left(a_{j+1,2k} + a_{j+1,2k+1} \right)$$

$$b_{jk} = \frac{1}{\sqrt{2}} \left(a_{j+1,2k} - a_{j+1,2k+1} \right).$$

We can say that the filter bank has the logarithmic tree structure since at each step of the decomposition we pass from a finer level j to a coarser level $j - 1$ with half as many outputs.

Any linear transform of a given vector can be represented as a multiplication of this vector by the transform matrix. In Section 6.1 we showed that any linear filtering can be interpreted as a linear transform with the constant-diagonal matrix (6.2). On the other hand, entries of a discrete transform matrix are samples of the basis functions. Now consider how hierarchical wavelet filtering can be represented as a linear transform. Assume that the input sequence $x(n)$ has length $N = 8$, then in matrix form the DWT of the sequence $x = (x(0), x(1), \ldots, x(7))^{\mathrm{T}}$ based on the Haar wavelet can be written as

$$y = Tx$$

where

$$T = \begin{pmatrix} c^3 & c^3 & c^3 & c^3 & c^3 & c^3 & c^3 & c^3 \\ -c^3 & -c^3 & -c^3 & -c^3 & c^3 & c^3 & c^3 & c^3 \\ -c^2 & -c^2 & c^2 & c^2 & 0 & 0 & 0 & 0 \\ 0 & 0 & 0 & 0 & -c^2 & -c^2 & c^2 & c^2 \\ -c & c & 0 & 0 & 0 & 0 & 0 & 0 \\ 0 & 0 & -c & c & 0 & 0 & 0 & 0 \\ 0 & 0 & 0 & 0 & -c & c & 0 & 0 \\ 0 & 0 & 0 & 0 & 0 & 0 & -c & c \end{pmatrix}$$

is the matrix with rows equal to the Haar wavelets, or

$$T = \begin{pmatrix} 1 & 1 & 1 & 1 & 1 & 1 & 1 & 1 \\ -1 & -1 & -1 & -1 & 1 & 1 & 1 & 1 \\ -1 & -1 & 1 & 1 & 0 & 0 & 0 & 0 \\ 0 & 0 & 0 & 0 & -1 & -1 & 1 & 1 \\ -1 & 1 & 0 & 0 & 0 & 0 & 0 & 0 \\ 0 & 0 & -1 & 1 & 0 & 0 & 0 & 0 \\ 0 & 0 & 0 & 0 & -1 & 1 & 0 & 0 \\ 0 & 0 & 0 & 0 & 0 & 0 & -1 & 1 \end{pmatrix}$$

scaled by $c = \frac{1}{\sqrt{2}}$. The scaling reflects the decimation process and gives unit vectors in the columns and rows, because $2c^2 = 1$, $4c^4 = 1$, and $8c^6 = 1$. It is easy to see that T describes three levels of wavelet decomposition. Its rows correspond to four fine differences, two intermediate differences, one coarse difference, and the overall average. In other words, T contains samples of Haar's wavelet and its six scaled and translated versions and samples of the box function (6.38). Since the wavelet transform based on the Haar wavelets is orthonormal, the inverse transform matrix used for synthesis represents the transposed matrix T.

One more interpretation of hierarchical filtering by Haar's filter bank can be given from the space decomposition point of view. For orthonormal bases (6.28) can be rewritten as

$$V_{j+1} = V_j \oplus W_j$$

where the symbol $\oplus$ denotes the orthogonal sum. Every function in V_{j+1} is a sum of the functions in V_j and W_j and the spaces V_j and W_j intersect only in the zero function.

It follows from the dilation equation (6.37) that

$$\varphi_{j-1,k}(t) = \frac{1}{\sqrt{2}} \left[\varphi_{j,2k}(t) + \varphi_{j,2k+1}(t) \right]$$

and the wavelet equation (6.39) gives

$$\psi_{j-1,k} = \frac{1}{\sqrt{2}} \left[\varphi_{j-1,2k}(t) - \varphi_{j,2k+1}(t) \right].$$

Thus, any function from V_{j-1} can be represented as $\sum_k a_{j-1,k}\varphi_{j-1,k}$ where

$$a_{j-1,k} = \frac{1}{\sqrt{2}} \left(a_{j,2k} + a_{j,2k+1} \right)$$

and any function from W_{j-1} can be represented as $\sum_k b_{j-1,k}\psi_{j-1,k}$ where

$$b_{j-1,k} = \frac{1}{\sqrt{2}} \left(a_{j,2k} - a_{j,2k+1} \right).$$

A function at the fine resolution j is equal to a combination of "average plus detail" at the coarse resolution $j - 1$,

$$\sum_k a_{jk}\varphi_{jk}(t) = \sum_k a_{j-1,k}\varphi_{j-1,k}(t) + \sum_k b_{j-1,k}\psi_{j-1,k}(t).$$

Now assume that the largest resolution level j is equal to 3 and consider the approximation $v_3(t)$, then the wavelet filtering corresponds to the three decomposition levels of the space V_3. First we represent V_3 as

$$V_3 = V_2 \oplus W_2.$$

At the second step, by decomposing V_2 we obtain

$$V_3 = V_1 \oplus W_1 \oplus W_2.$$

Finally, decomposing V_1 we obtain

$$V_3 = V_0 \oplus W_0 \oplus W_1 \oplus W_2.$$

The subspace V_3 contains all combinations of the piecewise constant functions on the intervals of length $\frac{1}{8}$. At each step of decomposition on the right side of the above equations the same space of functions is expressed differently. The functions in V_0 are constant on $(0, 1]$. The functions in W_0, W_1, and W_2 are combinations of wavelets with different resolutions. The function $v_3(t)$ in V_3 has a piece $w_j(t)$ in each wavelet subspace W_j plus a piece $v_0(t)$ in V_0,

$$v_3(t) = \sum_k a_{0k}\varphi_{0k}(t) + \sum_k b_{0k}\psi_{0k}(t) + \sum_k b_{1k}\psi_{1k}(t) + \sum_k b_{2k}\psi_{2k}(t).$$

6.7   Historical background

The first wavelet basis was described by Alfred Haar at the start of the twentieth century. This wavelet basis is named after him. In the 1930s, physicist Paul Levy applied the Haar wavelets to investigate Brownian motion. He found that the scale-varying basis functions were superior to the Fourier basis functions while dealing with the small details in Brownian motion. In the early 1980s the word *wavelet* appeared, owing to the physicist Alex Grossman and the geophysicist Jean Morlet. They used the French word *ondelette* which means "small wave." Later, this word was transformed into "wavelet" by translating "onde" into "wave." In the 1980s Stephane Mallat developed the multi-resolution theory. He showed a connection between wavelets, scaling functions, and filter banks (Mallat 1989). Ingrid Daubechies used this approach to construct probably the most important families of wavelets (Daubechies 1988; Cohen *et al.* 1992). They are compactly supported; that is, they lead to FIR filter banks. Unlike the Haar wavelets these families are based on rather smooth wavelet functions that make the obtained bases more suitable for multimedia applications (see, for example, (Antonini *et al.* 1992)).

6.8   Application of wavelet filtering to image compression

Consider how wavelet filtering can be used in order to perform the L-level wavelet decomposition of the image of the size $M \times N$ pixels (to be more precise, usually we decompose one of the so-called Y, U, or V components of the original image or a matrix of size $M \times N$ samples; for details, see Chapter 8). It is evident that the 2D wavelet decomposition is a separable transform. Thus, first we perform the wavelet filtering over the matrix rows and then the obtained matrix is filtered over the columns. At the first level of the wavelet hierarchical decomposition, the image is decomposed using two times downsampling (over the rows and over the columns) into high horizontal–high vertical (HH1), high horizontal–low vertical (HL1), low horizontal–high vertical (LH1), and low horizontal–low vertical (LL1) frequency subbands. They correspond to

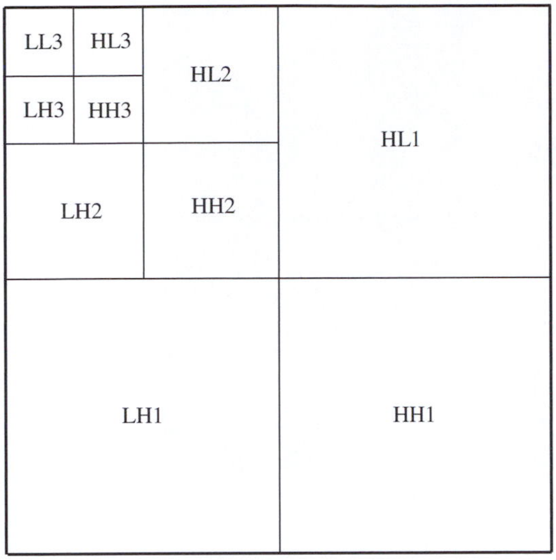

Figure 6.9 Wavelet decomposition of image

filtering by highpass filter $h_1(n)$ over rows and over columns, by highpass filter $h_1(n)$ over rows and lowpass filter $h_0(n)$ over columns, by lowpass filter $h_0(n)$ over rows and highpass filter $h_1(n)$ over columns, and by lowpass filter $h_0(n)$ over rows and columns, respectively. The LL1 subband is then further filtered and two times downsampled in order to produce a set of HH2, HL2, LH2, and LL2 subbands. This is done recursively L times to produce an array such as that illustrated in Fig. 6.9, where three downsamplings (over the rows and over the columns each) have been used. As a result we obtain $3L + 1$ matrices of reduced size.

In Fig. 6.11 the wavelet subbands obtained by the three-step decomposition of the gray-scale image "Bird"[1] of size 1000×1000 are shown. The original image is presented in Fig. 6.10. In total we have $3 \times 3 + 1 = 10$ matrices of wavelet coefficients. Most of the energy is in the low-lowpass subband LL3 of size 125×125. This upper left subimage is a coarse approximation of the original image. The other subbands add details. The high-frequency subbands HH1, HL1, and LH1 of size 500×500 contain mainly contours and sharp details. The high-frequency subbands of the next levels of decomposition contain many more details.

Each matrix is quantized by a scalar or vector quantizer and then encoded. The quantization step is chosen depending on the required compression ratio and bit allocation. Clearly, we can use larger steps when we quantize the subbands with low energy without a significant loss of the reconstructed image quality.

The quantized highpass subbands usually contain many zeros. They can be efficiently lossless encoded by using zero run-length coding followed by the Huffman coding of pairs (run length, amplitude) or by arithmetic coding. The lowpass subbands usually do not contain zeros at all or contain a small number of zeros and can be coded by the Huffman code or by the arithmetic code.

---

[1] The photo was taken by the author, December 2008 in Peterhoff, St Petersburg, Russia.

Figure 6.10   Original gray-scale image "Bird"

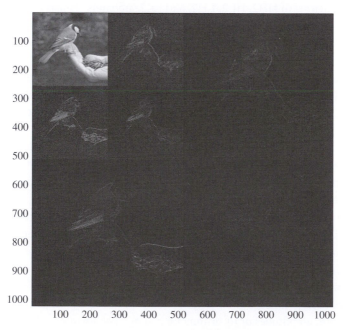

Figure 6.11   Wavelet subbands of gray-scale image "Bird"

6.9   Embedded zerotree coding and set partitioning in hierarchical trees

More advanced coding procedures used, for example, in the MPEG-4 standard try to take into account dependencies between subbands. One such method is called *zerotree coding* (Shapiro 1993). Figure 6.12 illustrates the *parent–child dependencies* of subbands. A single parent node in a subband of higher decomposition level has four child nodes in the corresponding subband of lower decomposition level. Each child node has four corresponding next-generation children nodes in the same type of subband of the next level. Parent–child dependencies between subbands are used to predict the insignificance of children nodes from the insignificance of their parent node. It is assumed that with a high probability the insignificant parent node is followed by the insignificant children nodes.

One approach typically used to lossless encode wavelet subband coefficients consists of encoding the so-called *significance map*, i.e. the binary decision whether a coefficient has zero or nonzero quantized value followed by encoding coefficient magnitudes. A rather large fraction of the bit budget is usually spent on encoding the significance map. In order to reduce the number of bits for significance map coding, the zerotree method implies the following classification of wavelet coefficients. A coefficient is said to be an element of a zerotree for the given threshold if it and all of its descendants (children) are insignificant with respect to this threshold. An element of a zerotree is a *zerotree root* if it is not the descendant of a previously found zerotree root; that is, it is not predictably insignificant from a zerotree root at a coarser scale. A zerotree root is encoded with a special symbol indicating that all its descendants at the finer scales are insignificant. Thus, the following four symbol types are used: zerotree root, isolated zero, which

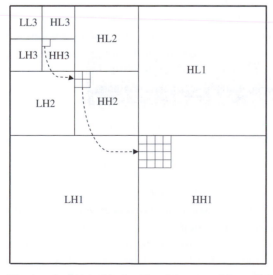

Figure 6.12   The wavelet hierarchical subband decomposition and the parent–child dependencies of the subband

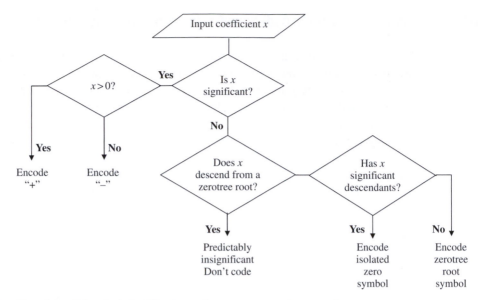

Figure 6.13   Flowchart of the pixel classification and zerotree generating procedure

means that the coefficient is insignificant but has some significant descendants, positive significant symbol, and negative significant symbol.

Figure 6.13 is a flowchart for encoding a coefficient of the significance map; or in other words, this figure illustrates how the zerotree based encoder classifies the wavelet coefficients and generates a zerotree. The main idea of the method is that we do not spend bits to encode descendants of the zerotree root.

Zerotree coding combined with *bit-plane* coding is called Embedded Zerotree Wavelet (EZW) coding (Shapiro 1993). A bit-plane is formed by bits of different wavelet coefficients which have the same significance. Coding schemes based on bit planes of wavelet coefficients produce embedded bitstreams which can be used for scalable image coding. It allows us quickly to reconstruct the low-resolution and low-quality image from a part of the embedded bitstream. By decoding more bits, a higher resolution and higher image quality can be obtained. Scalable image coding is discussed in detail in Chapter 8. Bit-plane coding also has the advantage of combining it with binary lossless coding schemes. In particular, it is often followed by binary arithmetic coding having rather low computational complexity (see Appendix).

The Set Partitioning In Hierarchical Trees (SPIHT) (Said and Pearlman 1996) algorithm was developed to provide lossless coding of wavelet coefficients as well as lossy scalable coding. This algorithm also exploits wavelet subband dependencies and bit-plane coding, and improves the performances of the EZW coding algorithm. In bit-plane based coding wavelet coefficients are successively compared with a system of predetermined decreasing thresholds. A given step of coding procedure consists of two passes called *sorting* and *refinement* passes. In the sorting pass, coefficients which are greater than or equal to the predetermined threshold (2^n) are classified as significant. Coordinates and signs of the significant coefficients are transmitted. Then, in the refinement pass the nth most significant bit of all coefficients which are greater than or equal to 2^{n+1} (their coordinates were transmitted in the previous sorting pass) are transmitted and the value n is decreased by 1. The algorithm can stop when the target rate or distortion is reached. In particular, it can be used for lossless compression.

In the SPIHT algorithm the described sorting pass is improved by taking into account parent–child dependencies of the wavelet subbands exploited by EZW coding. The algorithm is based on the so-called *spatial orientation trees*. The subband coefficients are grouped into sets corresponding to the spatial orientation trees. Then, the coefficients in each set are progressively coded from the most significant to the least significant bit-planes. The algorithm classifies not each coefficient but subsets of coefficients as significant if the maximum absolute value of the subset coefficient exceeds a given threshold and classifies it as insignificant otherwise. If a subset is encoded as insignificant, then a zero is transmitted; otherwise a one is transmitted. The subsets classified as significant are further partitioned into new subsets and the significance test is then applied to them. The procedure is repeated until each subset contains a single coefficient. The main idea is to partition coefficients into subsets in such a way that insignificant subsets would contain a large number of coefficients. It reduces the number of computations and the number of transmitted bits.

To explain the subset classification procedure we consider the spatial orientation trees shown in Fig. 6.14. Let $\mathcal{S}_R$ be a set of coordinates of tree roots which are the nodes at the lowest resolution level. Denote as $\mathcal{S}_D(i, j)$ the set of coordinates of all descendants (children) and as $\mathcal{S}_O(i, j)$ the set of coordinates of the direct descendants (offsprings) of the node with coordinates (i, j). Let $\mathcal{S}_L(i, j) = \mathcal{S}_D(i, j) \setminus \mathcal{S}_O(i, j)$ denote the set of coordinates of the node (i, j) descendants except the coordinates of the direct descendants. It is easy to see that each node either has no offsprings or has at least four (2×2) offsprings.

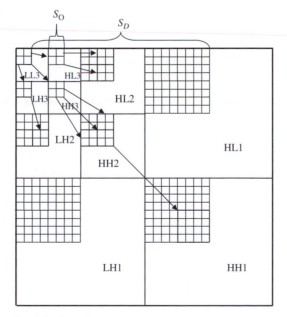

Figure 6.14   Spatial orientation trees

The indices of all coefficients are put into three lists: the List of Insignificant Coefficients (LIC), the List of Significant Coefficients (LSC), and the List of Insignificant Sets of coefficients (LIS). To initialize the algorithm we set LSC to the empty set and put the tree roots into LIC, and their sets $\mathcal{S}_D$ into LIS. Then we first check the significance of coefficients in LIC and then the significance of sets in LIS.

The following partitioning rules are used for splitting significant sets :

1. If the significant set is $\mathcal{S}_D(i, j)$, it is split into $\mathcal{S}_L(i, j)$ and the four coefficients which are the offsprings of the coefficient with coordinates (i, j).
2. If the significant set is $\mathcal{S}_L(i, j)$, it is split into the four $\mathcal{S}_D(k, l)$ sets which are the sets of descendants of the offsprings of the coefficient with coordinates (i, j), i.e. $(k, l) \in \mathcal{S}_O(i, j)$.

In the sorting pass, those coefficients from LIC which are found to be significant are moved to the LSC. For each of the significant coefficients, a "1" followed by the coefficient sign is transmitted. The significance of the sets from LIS is specified by one bit. If a set is insignificant, then a zero is transmitted; otherwise a one is transmitted. Those sets from LIS which are found to be significant are split into subsets according to the described rules and the new subsets are added to LIS if they are not empty. The refinement pass coincides with that for bit-plane coding. For those coefficients which are in the LSC except the coefficients found in the last sorting pass, the nth most significant bit is transmitted.

We considered two methods of lossy coding based on wavelet transform: EZW and SPIHT. In Chapter 8 we will describe the JPEG-2000 standard which is based on similar ideas. After applying wavelet decomposition to an input image, its energy localized in the low horizontal–low vertical wavelet subband of the highest (lowest resolution) decomposition level, correlation between coefficients inside high-frequency subbands as well as between different high-frequency subbands reduces compared to the correlation between pixels of the original image. However, some intra- and inter-subband correlation still exists after decomposition. The described wavelet-based coding techniques as well as the coding techniques used in the JPEG-2000 exploit this correlation and have the following common features:

- They are based on bit-plane coding that reduces the computational complexity and provides scalability of the encoded bitstream.
- They classify subband coefficients into sets taking into account intra- (JPEG-2000) or inter- (EZW and SPIHT) correlation between subband coefficients. This classification results in saving a number of bits for representation of the all-zero blocks of coefficients.

Problems

6.1   Does the filter with the pulse response

$$h(n) = (-1, 2, 6, 2, -1)$$

have linear phase?

6.2   Are the following filters invertible:

$$(a)\, h(n) = \left(\frac{3}{4}, -\frac{1}{2}\right)$$

$$(b)\, h(n) = (1/2, 1)?$$

6.3   Find the amplitude and the phase of the filter with the pulse response

$$h(n) = (-2, 4, -2).$$

Is it a lowpass or highpass filter?

6.4   Decimate the input sequence:

$$23, 4, 5, -1, 6, 7, 4, 8.$$

Upsample the decimated sequence.

6.5   Compute the decimated subband outputs for the Haar filter bank with input

$$x(n) = 0, 1, -1, 2, 5, 1, 7, 0.$$

Reconstruct the signal by feeding the upsampled subband components into the corresponding synthesis filter bank. Verify that the output is $x(n-1)$.

6.6   For the given analysis filter bank:

$$h_0(n) = (-1, 1, 8, 8, 1, -1)16/\sqrt{2}$$

$$h_1(n) = (-1, 1)/\sqrt{2}$$

find the frequency response of the lowpass and highpass filters. For the given input sequence

$$x(n) = (1, 1, 2, 3, 1, 5, 2)$$

check whether or not the filter bank

$$g_0(n) = (1, 1)/\sqrt{2}$$

$$g_1(n) = (-1, -1, 8, -8, 1, 1)16/\sqrt{2}$$

can be the inverse filter bank with respect to $(h_0(n), h_1(n))$.

6.7   Consider the analysis filter bank:

$$h_0(n) = (2, 8, 4, -1)/\sqrt{85}$$

$$h_1(n) = (1, 4, -8, 2)/\sqrt{85}$$

and the synthesis wavelet filter bank:

$$g_0(n) = (-1, 4, 8, 2)/\sqrt{85}$$

$$g_1(n) = (2, -8, 4, 1)/\sqrt{85}.$$

Assume that the length of the input sequence is $N = 6$. Find the frequency response of the lowpass and highpass filters. Write down the transform matrix of size $N \times N$ corresponding to the lowpass filter. Write down the transform matrix of size $N \times N$ corresponding to the highpass analysis filter.

6.8 For the filter banks described in Problem 6.7, write down the transform matrix of size $N \times N$ which describes the forward transform corresponding to the analysis filter bank (take decimation into account). Write down the transform matrix of size $N \times N$ which describes the inverse transform corresponding to the synthesis filter bank (take decimation into account). Show that the filter bank $(h_0(n), h_1(n))$ determines the invertible orthonormal transform.

6.9 Assume that the input sequence of the analysis filter bank described in Problem 6.7. is

$$x(n) = (5, 3, 10, 2, -1, -4).$$

Find low-frequency and high-frequency parts of the input sequence by using the analysis filter bank. Decimate the obtained lowpass and highpass filtered parts. Reconstruct the input sequence by using the synthesis filter bank described in Problem 6.7.

7 Speech coding: techniques and standards

In this chapter we consider first some of the requirements which different types of multimedia application can impose on speech coders. Then a brief overview of the existing speech coding standards will be given. In Chapter 4, we considered linear prediction techniques which are, in fact, used by almost all modern speech codecs. However, in addition to the high predictability, speech signals have specific features which are used by speech coders and allow us to achieve very high compression ratios. These properties are characterized by such parameters as *pitch period*, *linear spectral parameters*, etc. We will explain the meanings of these parameters and describe standard algorithms for finding them.

The most important attributes of speech coding are: bit-rate, delay, complexity, and quality.

The *bit-rate* is determined as the number of bits required in order to transmit or store one second of the speech signal. It is measured in kb/s. The range of bit-rates that have been standardized is from 2.4 kb/s for secure telephony up to 64 kb/s for network applications. Table 7.1 shows a list of standardized speech coders. It includes their bit-rates, delays measured in *frame size* and *lookahead* requirements, and their complexity in Million Instructions Per Second (MIPS). Several coders standardized by the International Telecommunication Union (ITU) are presented. For completeness some digital cellular and secure telephony standards are also included in Table 7.1.

The 64-kb/s G.711 Pulse Code Modulation (PCM) coder (G.711 1972) is used in digital telephone networks. The ITU standard G.726 (G.726 1990) describes a speech codec based on Adaptive Differential Pulse Code Modulation (ADPCM) operating at bit-rates of 16–40 kb/s. The most commonly used mode is 32 kb/s since this is half the rate of G.711. The 32-kb/s G.726 ADPCM coder doubles the usable capacity of digital telephone networks compared to the G.711 PCM coder. Other modes of G.726 are used, for example, in digital wireless phones.

The 64/56/48-kb/s G.722 and 16-kb/s G.728 coders are used in video teleconferencing over Integrated Services Digital Networks (ISDN) including one or several digital telephone channels with transmitting rate equal to 64 kb/s. Notice that in these services unipolar {0, 1} digital data from terminals and computers are converted into a modified bipolar {−1, 1} digital format. Then this bipolar data stream is transmitted to its destination. At its destination the data are converted back to their original unipolar format.

In G.722 the speech signal is sampled at 16 kHz. This is a wideband coder that can handle speech and audio signal bandwidths up to 8 kHz compared to 4 kHz

Table 7.1 ITU, cellular, and secure telephony speech-coding standards

Standard	Bit-rates	Frame size/lookahead	Complexity (MIPS)
ITU standards			
G.711 PCM	64 kb/s	0/0	0.01
G.726, G.727, ADPCM	16, 24, 32, 40 kb/s	0.125 ms/0	2
G.722 Wideband coder	48, 56, 64 kb/s	0.125/1.5 ms	5
G.728 LD-CELP	16 kb/s	0.625 ms/0	19
G.729 CS-ACELP	8 kb/s	10/5 ms	20
G.723.1 MPC-MLQ	5.3, 6.4 kb/s	30/7.5 ms	16
G.729 CS-ACELP annex A	8 kb/s	10/5 ms	11
Cellular standards			
RPE-LTP (GSM)	13 kb/s	20 ms/0	6
IS-54 VSELP	7.95 kb/s	20/5 ms	13.5
PDC VSELP (Japan)	6.7 kb/s	20/5 ms	8
IS-96 QCELP	8, 4, 2, 0.8 kb/s	20/5 ms	23/21/17/11
PDC PSI-CELP (Japan)	3.45 kb/s	40/10 ms	40
US secure telephony standards			
FS-1015 LPC-10E	2.4 kb/s	22.5/90 ms	8
FS-1016 CELP	4.8 kb/s	30/30 ms	20
MELP	2.4 kb/s	22.5/20 ms	21

in narrowband speech coders. The G.722 coder is based on the principle of Sub Band-ADPCM (SB-ADPCM). The signal is split into two subbands, and samples from both bands are coded using ADPCM techniques. G.728 is a low-delay coder. It is used in digital telephony over packet networks, especially Voice over Cable (VoCable) and Voice-over-Internet Protocol (VoIP) telephony, where low delay is required.

The G.723 (G.723.1 1993) and G.729 coders have been standardized for low-bit-rate multimedia applications over telephone modems. G.723, also known as G.723.1 (G.721 + G.723 combined) in more precise terms, is a standard speech codec providing two bit-rates: 5.3 and 6.4 kb/s. It was designed for video conferencing and telephony over POTS networks and is optimized for real-time encoding and decoding. The standard G.723.1 is a part of the H.323 (IP-telephony) and H.324 (POTS networks) standards for video conferencing.

The second key attribute of speech coders is the *delay*. Speech coders for real-time conversations cannot have too much delay otherwise they will be unsuitable for network applications. It is known that to maintain the conversation dynamics in the call, it is necessary to provide a delay below 150 ms. On the other hand, for multimedia storage applications the coder can have virtually unlimited delay and still be suitable for the application.

As follows from Table 7.1 the highest rate speech coders, such as G.711 PCM and G.726 ADPCM, have the lowest delay. To achieve higher degrees of compression, speech has to be divided into blocks or frames and then encoded frame by frame. For G.728 the frames are five samples (0.625 ms) long. For the first-generation cellular coders, the frames are 20 ms long. However, the total delay of a speech compression system includes the frame size, the look-ahead, and other algorithmic delays, processing delay and so on. The algorithm used for the speech coder determines the frame size and the look-ahead; that is, a need to use one or a few future frames in processing of the current frame. The complexity of speech coding influences the processing delay. The network connection determines the multiplexing and transmission delays.

In the past, speech coders were only implemented on Digital Signal Processing (DSP) chips or on other special-purpose hardware. Recent multimedia speech coders, however, have been implemented on the host Central Processing Unit (CPU) of personal computers. The measures of *complexity* for a DSP and a CPU are somewhat different due to the natures of these two systems. DSP chips from different vendors have different architectures and consequently different efficiencies in implementing the same coder. However, for fixed-point arithmetic processors such as, for example, ARM7, TMS320, Intel Xscale, etc., for most speech-coding algorithms, the number of MIPS in the optimized software implementation is approximately the same. Thus, the numbers given in Table 7.1 are approximate DSP MIPS for each coder, and they range from zero to 30 MIPS for all the ITU standard-based coders.

Quality is the speech-coder attribute that can be evaluated in a different way depending on the application. When speech coding was only used for secure telephony, quality was synonymous with intelligibility. For network applications, high intelligibility was assumed, and the goal was to preserve the naturalness or so-called subjective quality of the original speech. The high-rate speech coders operate on a sample-by-sample basis. Their quality is more or less directly related to the Signal-to-Noise Ratio (SNR) achieved when the speech samples are quantized

$$SNR = 10 \log_{10} \frac{E}{D} \text{dB}$$

where $E = \sum_{i=1}^{N} x_i^2$ denotes the energy of the original sequence $x = (x_1, x_2, \ldots, x_N)$ and $D = \sum_{i=1}^{N} (x_i - \hat{x}_i)^2$ is the energy of the quantization noise or, in other words, D represents the energy of the difference between the original x and the reconstructed $\hat{x}$ sequences.

At the low bit-rates used for secure telephony, however, speech coding is based on a *speech production model*. It typically results in worse quality of the synthesized speech generated by these coders. Figure 7.1[1] shows a plot of subjective speech quality for a

---

[1] The quality estimates presented as well as the complexity and performances in Table 7.1 were compiled from different sources.

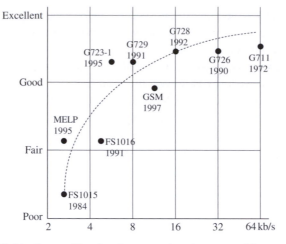

Figure 7.1   Subjective quality of various speech coders versus bit-rate

range of telephone bandwidth coders spanning bit-rates from 64 kb/s down to 2.4 kb/s. A five-point rating scale was used:

- 5: excellent quality, no noticeable distortions;
- 4: good quality, only very slight distortions;
- 3: fair quality, noticeable but acceptable distortions;
- 2: poor quality, strong distortions;
- 1: bad quality, highly degraded speech.

It can be seen from Fig. 7.1 that telephone bandwidth coders maintain a uniformly high quality for bit-rates ranging from 64 kb/s down to about 8 kb/s but the subjective quality goes down for bit-rates below 8 kb/s. However, recent implementations of the G.723.1 standard have rather good subjective quality. Typically, vocoders have relatively low subjective quality. For all speech coders providing bit-rates below 3 kb/s there are at least some noticeable distortions in the synthesized speech. The main requirement in this range of bit-rates is high intelligibility. In 1995 the Mixed Excitation Linear Prediction (MELP) coder was standardized. This is the United States Department of Defense (DoD) speech coding system. It is used in military applications, satellite communications, secure voice and secure radio devices. The MELP vocoder provides bit-rate 2.4 kb/s with surprisingly good quality.

The two main speech signal features which are exploited by most standard codes are:

- narrowness of band (frequency band from 50 Hz up to 5 kHz);
- quasi-stationarity (the speech signal can be split into short blocks (frames), each of which can be regarded as a random stationary process).

Now we consider some speech coding standards in more detail. Standards play a major role in the multimedia revolution because they provide interoperability between hardware and software developed by different companies. Although, in practice, one could use a nonstandard coder, it is essential that at least one standard speech coder be mandatorily implemented in the developed soft- and hardware. This assures that voice communications are always initiated and maintained. Additional optional speech coders can be included at the developer's discretion.

Nowadays, all standard speech codecs can be split into four large classes depending on the algorithms on which they are based:

- coding based on PCM (G.711);
- linear predictive coding based on ADPCM (G.726, G.727, G.722);
- Linear Predictive Analysis-by-Synthesis coding (LPC-AS) (G.723.1, G729, G.728, IS-54, GSM);
- LPC vocoders (MELP).

7.1   Direct sample-by-sample quantization: Standard G.711

The G.711 PCM coder is designed for telephone bandwidth speech signals. Its input signal is a discrete-time speech signal sampled with rate 8 kHz. The coder performs direct sample-by-sample scalar quantization. Instead of uniform quantization, a form of nonuniform quantization known as *companding* is used. The name is derived from the words "compressing-expanding." The idea behind this method is illustrated in Fig. 7.2. A sample x of the original speech signal is compressed using a memoryless nonlinear device described by function $F(x)$. The sample y thus obtained is then uniformly quantized. The dequantized sample $\hat{y}$ is expanded using a nonlinear function $F^{-1}(x)$ which is the inverse of that used in compression.

Before quantization, the signal is distorted by a function similar to that shown in Fig. 7.3. This operation compresses the large values of the waveform while enhancing the small values. Strictly speaking, our goal is to compress those values which should be quantized with a coarser scale and enhance values which should be quantized with a finer scale. If the distorted signal is then uniformly scalar quantized, it is equivalent to a scalar quantization which starts with a small quantization step which then becomes larger for higher signal values. There are different standards for companding. North America and Japan use a standard compression curve known as μ-law companding. Europe has adopted a different, but similar, standard known as A-law companding.

The μ-law compression is described by

$$F(x) = F_{\max}\mathrm{sgn}(x)\frac{\log_2\left(1 + \mu\,|x/x_{\max}|\right)}{\log_2\left(1 + \mu\right)}$$

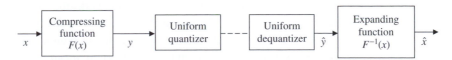

Figure 7.2   The concept of companding

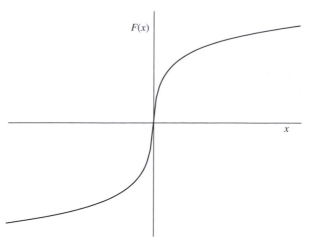

Figure 7.3 $\mu = 255$ compression curve

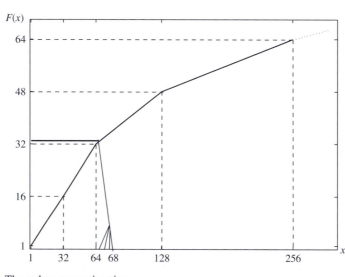

Figure 7.4   The μ-law approximation

where x is the input sample value, sgn(x) denotes the sign of x, and x_{max} and F_{max} are the maximum values of the input and output sample, respectively. The parameter μ, defines the degree of the curvature of the function. G.711 uses value $\mu = 255$.

Let us consider the implementation of the $\mu = 255$ encoder for 8-bit PCM coding from G.711. The $\mu = 255$ curve is approximated by a piecewise linear curve, as shown in Fig. 7.4. The curve is an odd function and we show the initial part of the positive input portion only. Notice that when using the μ-law in networks the suppression of the all-zero character signal is required. The positive portion of the curve is approximated by eight straight-line segments. We divide the positive output region into eight equal segments, which effectively divides the input region into eight *unequal* segments. Each output segment contains 16 samples; that is, the value of the sample in each segment is quantized using 4 bits. To specify the segment in which the sample lies, we spend 3 bits and 1 bit gives the sign of the sample: 1 for positive, 0 for negative. Thus, in total, 8 bits are spent for each sample.

For example, in order to quantize the value 66 we send the sign bit 1, the number of segment 010, and the number of quantization level 0000 (the values 65, 66, 67, 68 are quantized into 33). At the decoder using codeword 10100000 we reconstruct the approximating value $(64 + 68)/2 = 66$. Notice that the same approximating value will be reconstructed for the values 65, 66, 67, and 68.

Each segment of the input axis is twice as wide as the segment to its left. For example, the values 129, 130, 131, 132, 133, 134, 135, and 136 are quantized to the same value 49. Thus, the resolution of each segment is twice as bad as that of the previous one.

The G.711 PCM provides the lowest possible delay (a single sample) and provides the lowest possible complexity as well. It is a high-rate coder. Its bit-rate is 64 kb/s. G.711 is the default coder for ISDN telephony.

7.2 ADPCM coders: Standards G.726, G.727

This group of coders is based on the linear prediction method. As mentioned in Chapter 4 the main idea of this method is to predict the value of the current sample by a linear combination of previous already reconstructed samples and then to quantize the difference between the actual value and the predicted value. The linear prediction coefficients are the weighting coefficients used in the linear combination. Now we consider how this method can be applied to the compression of speech signals.

If the prediction is simply the last sample and the quantizer has only one bit, the system becomes a *delta-modulator*, illustrated in Fig. 4.6. *Delta modulation* is a simple technique for reducing the dynamic range of the numbers to be encoded. Instead of sending each sample value, we send the difference between a sample and a value of a staircase approximation. The approximation can only either increase or decrease by a step Δ at each sampling point. If the staircase value is below the input sample value, we increase the approximation by the step Δ; otherwise we decrease it by the step Δ. The output of the quantizer is "1" if the difference is positive and "0" otherwise. Figure 7.5 shows the analog waveform and the staircase approximation. The transmitted bitstream for the example shown in Fig. 7.5 is 1 1 1 1 1 0 0 0 . . .

In order to use the delta modulation approach efficiently it is necessary to choose the step size Δ and the sampling frequency properly. To generate the staircase approximation close to the original input signal the delta modulation requires sampling frequencies greater than the Nyquist frequency. Increasing the sampling frequency, however, means that the delta-modulated waveform requires a larger bandwidth.

In order to take into account the fastest possible change in the signal, the step size has to be increased. On the other hand, increasing the step size increases the quantization error.

Figures 7.6 and 7.7 show the consequences of the incorrect choice of step size. If the step is too small this circumstance can lead to a *slope overload* condition, where

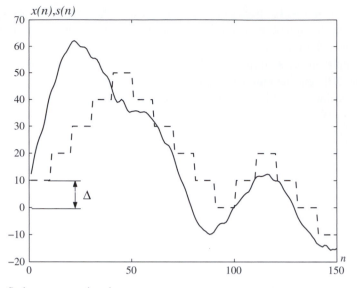

Figure 7.5 Staircase approximation

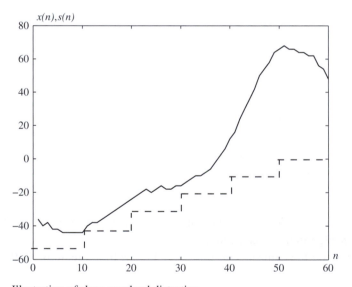

Figure 7.6 Illustration of slope overload distortion

the staircase cannot trace fast changes in the input signal. If, on the other hand, the step is too large, a considerable overshoot will occur during periods when the signal is changing slowly. In that case, we have significant distortions called *granular noise*.

We have shown that the appropriate step size used in delta modulation should depend on how fast the signal is changing from one sample to the next. When the signal is changing fast, a larger step size will reduce overload. When the signal is changing slowly, a smaller step size will reduce the amount of overshoot, and therefore the quantization error.

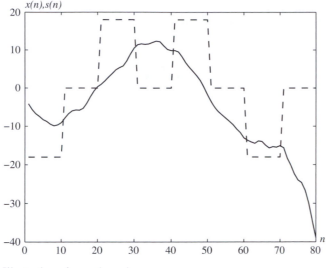

Figure 7.7 Illustration of granular noise

Adaptive delta modulation is a scheme which permits an adjustment of the step size depending on the performances of the input signal. It is important that the receiver is able to adapt the step sizes in exactly the same manner as the transmitter. If this is not the case, the receiver cannot uniquely recover the staircase approximation. All that is being transmitted is a bitstream and the step size can be derived from the received bitstream.

The idea of step size adaptation is very simple. If the bitstream contains almost equal numbers of 1's and 0's we can assume that the staircase approximation is oscillating about a slowly varying input signal. In such cases, we should reduce the step size. On the other hand, an excess of either 1's or 0's within a stream of bits would indicate that the staircase approximation is trying to catch up with the input signal. In such cases, we should increase the step size.

We shall now consider two adaptation rules, namely, the *Song algorithm* and the *Space Shuttle* algorithm (Roden 1988).

The Song algorithm compares the transmitted bit with the previous bit. If the two bits are the same, the step size is increased by a fixed amount δ. If the two bits are different, the step size is reduced by the fixed amount δ. However, if an input signal first changes fast and then suddenly begins to change rather slowly, the Song algorithm can lead to a damped oscillation with a rather large amplitude.

The Space Shuttle algorithm is a modification of the Song algorithm and it removes the damped oscillations. If the current bit is the same as the previous bit, the step size increases by a fixed amount δ. However, when the bits disagree, the step size immediately reduces to its minimum size Δ. This is the difference compared to the Song algorithm where the step size decreases towards zero at a rate of δ per sample. Coders based on delta modulation provide bit-rates approximately 2–4 times lower than that of the PCM coder. Nowadays, they are not used as standard coders.

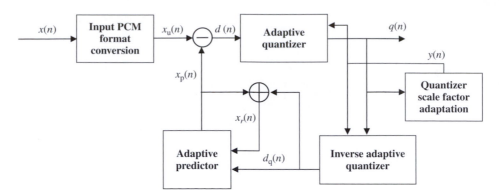

Figure 7.8   Standard G.726 speech coder

A coder based on Differential Pulse Code Modulation (DPCM) is shown in Fig. 4.5. It predicts a current sample value using a linear combination of several previous samples and then quantizes the difference between the actual value of the sample and the predicted value. As mentioned in Chapter 4, this coder does not exploit all the advantages of linear predictive coding. In particular, it uses nonoptimal linear prediction coefficients and the prediction is based on past reconstructed samples and not on true samples. Usually, the coefficients of the prediction are chosen by using some empirical rules and are not transmitted. For example, the quantizer shown in Fig. 4.5 uses, instead of actual values of prediction error $e(n)$, the values $d(n)$ reconstructed in the dequantizer (approximating values) and instead of the true sample values $x(n)$ their estimates $x_R(n)$ obtained via $d(n)$ which are not exactly equal to $e(n)$.

The speech coders G.726 and G.727 are ADPCM coders for telephone bandwidth speech. The coder scheme is shown in Fig. 7.8. The input signal is a 64 kb/s PCM speech signal (the sampling rate is 8 kHz and each sample is an 8-bit number). The first block converts the input sample $x(n)$ from the A-law or μ-law PCM to a uniform PCM sample $x_u(n)$. The second block calculates the difference $d(n)$ from the uniform PCM sample $x_u(n)$ and its predicted value $x_p(n)$:

$$d(n) = x_u(n) - x_p(n).$$

A nonuniform adaptive scalar quantizer with bit-rates 5, 4, 3, or 2 bits/sample is used to quantize the difference signal $d(n)$ for operating at 40, 32, 24, or 16 kb/s, respectively. Before quantization, $d(n)$ is scaled by $y(n)$ and converted to a logarithmic representation:

$$l(n) = \log_2 |d(n)| - y(n)$$

where $y(n)$ is computed by the scale factor adaptation block. The value $l(n)$ is then scalar quantized with a given quantization rate.

For example, to provide a bit-rate equal to 16 kb/s the absolute value $|l(n)|$ is quantized with $R = 1$ bit/sample and one more bit is used to specify the sign. The two quantization cells are: $(-\infty, 2.04)$ and $(2.04, \infty)$. They contain the approximating values 0.91 and 2.85, respectively.

The inverse adaptive quantizer reconstructs an approximation $d_q(n)$ of the difference signal using $y(n)$ and the corresponding approximating value for $l(n)$.

The adaptive predictor is used to compute the predicted value $x_p(n)$ from the reconstructed difference $d_q(n)$. More precisely a linear prediction is based on the two previous samples of the reconstructed signal $x_r(n)$ and the six previous values of the reconstructed difference $d_q(n)$. The predictor is described by the recurrent equation

$$x_p(n) = \sum_{i=1}^{2} a_i(n-1)x_r(n-i) + e(n)$$

where

$$e(n) = \sum_{i=1}^{6} b_i(n-1)d_q(n-i)$$

and

$$x_r(n) = x_p(n) + d_q(n)$$

are the reconstructed sample values, a_i and b_i are prediction coefficients.

The corresponding description of the predictor in the z-transform domain is

$$X_p(z) = \sum_{i=1}^{2} a_i X_r(z)z^{-i} + \sum_{i=1}^{6} b_i D(z)z^{-i}$$

where $X_p(z)$ denotes the z-transform of the sequence $x_p(n)$, $X_r(z)$ is the z-transform of the sequence $x_r(n)$, and $D(z)$ is the z-transform of the sequence $d_q(n)$. If we do not pay attention to the difference between $x(n)$ and $x_r(n)$ ($e(n)$ and $d_q(n)$), we obtain that the prediction error satisfies the following equation

$$E(z) = X(z) - X_p(z) = X(z)\left(1 - \sum_{i=1}^{2} a_i z^{-i}\right) - \sum_{i=1}^{6} b_i E(z)z^{-i}$$

where $X(z)$ is the z-transform of the sequence $x(n)$ and $E(z)$ is the z-transform of the sequence $e(n)$. Thus, the transfer function of the prediction filter has the form

$$H_e(z) = \frac{E(z)}{X(z)} = \frac{1 - \sum_{i=1}^{2} a_i z^{-i}}{1 + \sum_{i=1}^{6} b_i z^{-i}}.$$

The predictor coefficients a_i, b_i as well as the scale factor y are updated on a sample-by-sample basis in a *backward adaptive fashion*; that is, using the past reconstructed (synthesized) values. For example, for the second-order predictor:

$$a_1(n) = (1 - 2^{-8})a_1(n-1) + (3 \cdot 2^{-8})\text{sgn}(p(n))\text{sgn}(p(n-1))$$

$$a_2(n) = (1 - 2^{-7})a_2(n-1) + 2^{-7}\{\text{sgn}(p(n))\text{sgn}(p(n-2))\}$$

$$-2^{-7}f\{a_1(n-1)\}\,\text{sgn}(p(n))\text{sgn}(p(n-1))$$

where sgn(x) denotes the sign of x,

$$p(n) = d_q(n) + e(n)$$

and

$$f(a_1) = \begin{cases} 4a_1 & \text{if } |a_1| \le 2^{-1} \\ 2\text{sgn}(a_1) & \text{if } |a_1| > 2^{-1}. \end{cases} \tag{7.1}$$

For the sixth-order predictor:

$$b_i(n) = (1 - 2^{-8})b_i(n-1) + 2^{-7}\text{sgn}(d_q(n))\text{sgn}(d_q(n-i))$$

for $i = 1, 2, \ldots, 6$.

It is easy to see that the predictor coefficients are adjusted in a manner similar to the quantization step Δ in the adaptive delta modulation coder. If the signs of prediction errors at the steps n and $n - i$ coincide, the ith prediction coefficient increases; otherwise it decreases.

Both coders (G.726 and G.727) can operate using 2, 3, 4, or 5 bits/sample, corresponding to rates of 16, 24, 32, and 40 kb/s. The difference between these two coders is that G.727 uses embedded quantizers, while G.726 uses individually optimized quantizers for each of the four bit-rates. G.726 was created for circuit multiplication applications. In other words, it is used in multiplexing schemes which allow a group of voice-band channels to be compressed into a broadband channel. G.727 was created for packet circuit multiplication applications.

The embedded quantizer feature allows the least significant bits to be dropped if there is network overload.

7.3   Linear prediction analysis-by-synthesis coders (LPC-AS)

So far, the most popular class of speech coders for bit-rates between 4.8 and 16 kb/s is model-based coders which use a LPC-AS method. A linear prediction model of speech production (adaptive linear prediction filter) is excited by an appropriate excitation signal in order to model the signal over time. The parameters of both the filter and the excitation are estimated and updated at regular time intervals (frames). The compressed speech file contains these model parameters estimated for each frame.

Figure 7.9 shows a model of speech production. The linear prediction filter models those changes which occur in the *vocal tract* formed by the mouth, tongue, and teeth when we speak. A different shape of the vocal tract is influenced mainly by the tongue position. Roughly speaking, each sound corresponds to a set of filter coefficients. The vocal tract can be regarded as an acoustic tube of varying diameter which is characterized by resonance frequencies. This explains why the filter which models the vocal tract is often represented by the poles of its frequency response called *formant frequencies* or *formants*. The filter excitation depends on the type of the sound: voiced, unvoiced, vowel, hissing, or nasal. The voiced sounds are generated due to vibration of the vocal cords which interrupt the airstream from the lungs and produce a quasi-periodic impulse

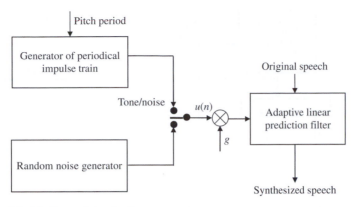

Figure 7.9 Model of speech production

train. The unvoiced signals are generated by noise-like signals. The simplest model shown in Fig. 7.9 uses two types of excitation signal (periodical impulse train and quasi-random noise) that are switched at each frame. The period of vocal cords' oscillation is called *pitch period* or *pitch*. It is estimated using the original speech signal and determines the period of impulses in the impulse train. We will consider forward adaptive and backward adaptive LPC-AS coders. In the forward adaptive LPC-AS coders the linear prediction is based on true past speech samples. The backward adaptive LPC-AS coders use reconstructed past samples in order to predict a current speech sample.

The main feature of LPC-AS coders is that in these coders a filtered excitation signal is compared with the original speech in order to find the best excitation for the linear predictive filter. As was shown in Chapter 4 the linear filtering can be interpreted as a linear transform but this transform is not an orthonormal transform. It means that we cannot search for the best excitation in the excitation (transform) domain. A prediction error (excitation) of small energy can lead to a large energy difference between the synthesized and original speech signals.

We start with the so-called CELP coder which is the basis of all LPC-AS coders (G.729, G.723.1, G.728, IS-54, IS-96, RPE-LTP (GSM), FS-1016 (CELP), etc.). CELP is an abbreviation of Code Excited Linear Predictive coder. CELP coding is a frame-oriented technique. The coder divides a sampled input speech signal into *frames*, i.e. blocks of samples, each of which is further divided into subframes. CELP coding is based on linear prediction, analysis-by-synthesis search procedures, and vector quantization with the Minimum Squared Perceptually Distorted Error (MSPDE) criterion. Notice that the perceptual distortion of the error signal improves the subjective quality of the synthesized speech by exploiting the masking properties of human hearing.

A CELP coder is shown in Fig. 7.10. It uses input signals at a sampling rate of 8 kHz and a 30 ms frame size with four 7.5 ms subframes. For each frame, a set of 10 coefficients of the Linear Prediction (LP) filter is determined by the so-called *open loop* linear prediction analysis on the input speech. The LP filter is used to model the *short-term* spectrum of the speech signal, or its format structure. The short-term linear prediction analysis is performed once per frame by autocorrelation analysis

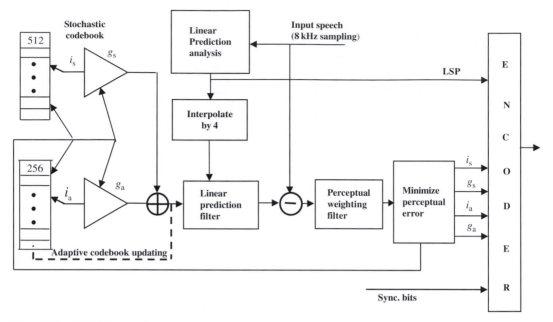

Figure 7.10   CELP transmitter

using a 30 ms Hamming window. The coefficients of the LP filter are found by using the Levinson–Durbin recursive procedure (see Chapter 4). Then the obtained *Linear Prediction Coefficients* (LPCs) are often converted into *linear spectral parameters* (LSPs), which are functions of the speech formants. The algorithm of converting LPCs to LSPs will be considered in Section 7.5. The LP filter is transmitted by its LSPs which are quantized by a nonuniform scalar quantizer. The total number of bits for representing 10 LSPs is 34. Since the LSPs are transmitted only once per frame but are needed for each subframe, they are linearly interpolated to form an intermediate set for each of the four subframes. The LP filter has a transfer function of the form $H_a(z) = 1/H_e(z)$.[2] The perceptual weighting filter, with the transfer function $H_e(z)/H_e(z/\gamma)$, $\gamma = 0.8$, is formed by a bandwidth expansion of the denominator filter using a weighting factor γ.

The excitation for the LP filter is calculated for each subframe as a sum of scaled entries from two codebooks. An *adaptive codebook* is used to predict the periodical (pitch) part of the excitation. It is often said that the adaptive codebook models the *long-term* speech periodicity. The MSPDE criterion is used for an adaptive codebook search which is often called a *closed loop* analysis-by-synthesis search. This analysis implies that in order to compute a prediction error, the filtered codeword is compared with the original speech. The method combines vector quantization of the excitation and the first-order linear prediction. The adaptive codebook contains 256 codewords. Since the two codebook searches are the most computationally consuming procedures in the CELP coder it is allowed to search any subset of the adaptive codebook. Each codeword of

---

[2] In speech-coding standards, the inverse of linear prediction filter is often called the linear prediction filter.

Table 7.2 Adaptive codebook

Index	Delay	Adaptive CB sample numbers
255	147	$-147, -146, -145, \ldots, -89, -88$
254	146	$-146, -145, -144, \ldots, -88, -87$
253	145	$-145, -144, -143, \ldots, -87, -86$
...	...	...
131	61	$-61, -60, -59, \ldots, -2, -1$
...	...	...
1	21	$-21, \ldots, -2, -21, \ldots, -2, -21, \ldots, -2$
0	20	$-20, \ldots, -1, -20, \ldots -1, -20, \ldots, -1$

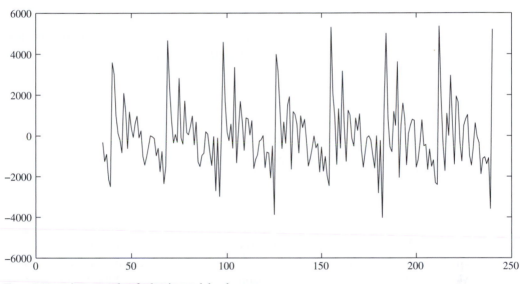

Figure 7.11   An example of adaptive codebook content

this book is constructed by repeating the previous excitation signal of length 60 samples delayed by $20 \leq M \leq 147$ samples. For delays less than the subframe length ($M < 60$) the codewords contain the initial M samples of the previous excitation vector. To complete the codeword to 60 elements, the short vector is replicated by a periodic extension. The indices of the previous excitation samples corresponding to the codewords of the adaptive codebook are shown in Table 7.2.

An example of the adaptive codebook content is shown in Fig. 7.11. The current subframe is shown in bold in the same figure. In order to find the best excitation in the adaptive codebook, first-order linear prediction is used. Let s be the perceptually distorted original speech vector and a_i be a filtered codeword c_i of the adaptive codebook; then we search for

$$\min_i \left\{ \|s - g_a a_i\|^2 \right\} = \min_i \left\{ \|s\|^2 - 2g_a(s, a_i) + g_a^2 \|a_i\|^2 \right\} \qquad (7.2)$$

where g_a is a prediction coefficient or the *adaptive codebook gain*.

By taking the derivative of the right part of (7.2) with respect to g_a and setting it to zero, we find that the optimal gain is

$$g_a = \frac{(s, a_i)}{\|a_i\|^2}.$$

In other words, the optimal gain is the ratio of the cross-correlation of the original speech and the filtered codeword to the energy of the filtered codeword. Inserting the optimal gain value into (7.2), we obtain

$$\min_i \left\{ \|s - g_a a_i\|^2 \right\} = \min_i \left\{ \|s\|^2 - \frac{2(s, a_i)^2}{\|a_i\|^2} + \frac{(s, a_i)^2}{\|a_i\|^2} \right\}$$

$$= \min_i \left\{ \|s\|^2 - \frac{(s, a_i)^2}{\|a_i\|^2} \right\}. \tag{7.3}$$

Minimizing (7.3) over i is equivalent to maximizing the last term in (7.3), since the first term is independent of the codeword a_i. Thus, the adaptive codebook search procedure finds a codeword c_i which maximizes the so-called *match function* m_i:

$$m_i = \frac{(s, a_i)^2}{\|a_i\|^2}.$$

The adaptive codebook index i_a and gain g_a are transmitted four times per frame (every 7.5 ms). The gain is coded between -1 and $+2$ using nonuniform, scalar, 5-bit quantization.

The residual between the filtered excitation from the adaptive codebook and the original speech signal is vector quantized using a fixed *stochastic codebook*. The search procedures for the stochastic and adaptive codebooks are virtually identical, differing only in their books and target vectors. The target for the adaptive codebook search is the perceptually distorted original speech. The stochastic codebook search target is the difference between the perceptually distorted original speech and the filtered adaptive codebook excitation; that is,

$$u = s - g_a a_{opt}.$$

The stochastic codebook search is performed by closed-loop analysis using the MSPDE criterion. We find such a codeword x_i which maximizes the following match function

$$\frac{(u, y_i)^2}{\|y_i\|^2}$$

where y_i is the filtered codeword x_i. The corresponding gain value is determined as

$$g_s = \frac{(u, y_i)}{\|y_i\|^2}.$$

The stochastic codebook contains 512 codewords. To reduce the computational complexity it is allowed to search any subset of this codebook. The codebook index i_s

and gain g_s are transmitted four times per frame. The gain (positive and negative) is coded using 5-bit, nonuniform scalar quantization. A special form of a stochastic codebook containing ternary valued samples $(-1, 0, +1)$ is used. This book is obtained by ternary level quantizing sequences of zero-mean, unit variance Gaussian samples.

The adaptive codebook content is updated by the scaled sum of the found excitations; that is,

$$g_a c_a + g_s x_s.$$

In other words, this scaled sum is moved into the adaptive codebook and the previous content is shifted out in such a way that the most distant 60 samples will be lost.

Thus, the analysis consists of three basic steps: (1) short-term prediction, (2) long-term adaptive codebook search, and (3) innovation stochastic codebook search. The CELP synthesis consists of the corresponding three synthesis steps performed in reverse order. The transmitted CELP parameters are the stochastic codebook index and gain, the adaptive codebook index and gain, and 10 LSPs which are computed from the obtained 10 coefficients of the LP filter.

The total number of bits per frame can be computed as

$$4(b_{g_s} + b_{i_s} + b_{g_a} + b_{i_a}) + b_{LSP} = 4(5 + 8 + 5 + 9) + 34 = 142$$

where b_{g_s}, b_{i_s} and b_{g_a}, b_{i_a} are numbers of bits for index and gain of the stochastic and adaptive codebook, respectively, b_{LSP} is the number of bits for the LSPs. Taking into account that the frame duration is 30 ms, we obtain that the bit-rate of the CELP coder is

$$R = \frac{142}{30 \cdot 10^{-3}} \approx 4733 \text{ b/s.}$$

Adding bits for synchronization and correcting errors, we obtain the bit-rate 4.8 kb/s.

Consider forward adaptive LPC-AS coders – 8 kb/s G.729 coder, and 6.4 and 5.3 kb/s G.723.1 coder. These coders are based on the same principles as the CELP coder; that is, on the principles of linear prediction analysis-by-synthesis coding and attempt to minimize a perceptually weighted error signal. However, they have some distinguishing features.

The G.723.1 coder operates on frames of 240 samples each or of duration 30 ms at an 8 kHz sampling rate. Each block is divided into four subframes, of 60 samples each. For every subframe, a 10th-order LP filter is computed using the original speech signal. The coefficients of the LP filter (LPCs) are transformed to the LSPs. Only the set of LSPs for the last subframe is quantized using a vector quantizer and stored or transmitted. Both the quantized and the unquantized LPCs are used to construct the short-term perceptual weighting filter, which is used to filter the entire frame and to obtain the perceptually distorted speech signal. The vector quantization uses three codebooks of 256 codewords each. By using the vector quantization, only 24 bits are required to transmit the filter parameters instead of the 34 bits spent by the CELP encoder.

Since the two codebook searches in CELP are extremely computationally consuming procedures, the G.723.1 standard exploits another approach in order to approximate periodical and nonperiodical parts of the speech signal. The adaptive codebook search is replaced by a pitch prediction. First, for every two subframes (120 samples) the open loop pitch period p is estimated by using the perceptually distorted speech signal $s(n)$. The pitch period is searched in the range from 18 to 145 samples; that is,

$$p = \arg\left\{ \max_{18 \leq j \leq 145} \frac{\left(\sum_{i=0}^{119} s(n+i)s(n+i-j)\right)^2}{\sum_{i=0}^{119} s(n+i-j)^2} \right\}.$$

Then the speech is processed on a 60 samples subframe basis. Using the pitch period estimate, p, and the pulse response of the fifth-order pitch prediction filter $a = (a_{-2}, a_{-1}, \ldots, a_2)$, a closed loop pitch prediction is computed by

$$\min_{j,a}\left\{ \sum_{n=0}^{59}\left(s(n) - \sum_{i=-2}^{i=2} a_i s(n - p + i - j) \right)^2 \right\}.$$

For subframes 0 and 2 the closed loop pitch lag j is selected from around the found open loop pitch lag p in the range ± 1 and $p + j$ is coded using 7 bits. (Note that the open loop pitch lag is never transmitted.) For subframes 1 and 3 the closed loop pitch lag is coded differentially using 2 bits and may differ from the previous subframe lag only by -1, 0, $+1$ or $+2$. The vector a of pitch predictor gains is vector quantized using two codebooks with 85 or 170 entries for the high bit-rate and 170 entries for the low bit-rate. The 170-entry codebook is the same for both rates. For the high rate if the open loop pitch is less than 58, then the 85-entry codebook is used for the pitch gain quantization. Otherwise, the pitch gain is quantized using the 170-entry codebook. The periodical (pitch) part of the speech approximation

$$s_{\mathrm{p}}(n) = \sum_{i=-2}^{2} a_i^{\mathrm{opt}} s(n - p + i - j^{\mathrm{opt}}), n = 0, 1, \ldots, 59$$

plays the same role as a filtered excitation from the adaptive codebook in the CELP standard. It is subtracted from the target vector $s(n), n = 0, 1, \ldots, 59$, to obtain the residual signal $r(n), n = 0, 1, \ldots, 59$.

Finally, the nonperiodic component of the excitation is approximated using the obtained residual signal. For the high bit-rate, a Multi-Pulse Maximum-Likelihood Quantized (MP-MLQ) excitation is used, and for the low bit-rate, an Algebraic Codebook Excitation (ACELP) is used.

The residual signal $r(n), n = 0, 1, \ldots, 59$, is approximated by

$$\hat{r}(n) = \sum_{j=0}^{n} h(j)v(n - j), 0 \leq n \leq 59$$

where $v(n)$ is the excitation for the filter with pulse response $h(n)$ and is determined as

$$v(n) = G \sum_{m=0}^{P-1} \beta_m \delta(n - l_m), 0 \leq n \leq 59$$

Table 7.3 ACELP pulse positions

Amplitude	Positions
±1	0, 8, 16, 24, 32, 40, 48, 56
±1	2, 10, 18, 26, 34, 42, 50, 58
±1	4, 12, 20, 28, 36, 44, 52, (60)
±1	6, 14, 22, 30, 38, 46, 54, (62)

where G is the gain factor, $\delta(n)$ is Kronecker's delta function, β_m and l_m are the amplitudes (± 1) and positions of the delta functions, respectively, and P is the number of pulses, which is 6 for even subframes and 5 for odd subframes. The pulse positions can be either all odd or all even. This is specified by one bit. Thus, the problem of finding the best approximation of the nonperiodic part is reduced to the problem of estimating the unknown parameters G, β_m, l_m, $m = 0, 1, \ldots, P - 1$, which minimize the squared error

$$\sum_{n=0}^{59} \left(r(n) - G \sum_{m=0}^{P-1} \beta_m h(n - l_m) \right)^2. \tag{7.4}$$

The following gain estimate is computed

$$G_{\max} = \frac{\max_{j=0,1,\ldots,59} \left\{ \left| \sum_{n=j}^{59} r(n) h(n - j) \right| \right\}}{\sum_{n=0}^{59} h(n)^2}.$$

The estimated gain $G_{\max}$ is scalar quantized by the nonuniform scalar quantizer with 24 cells. Each cell is equal to 3.2 dB, i.e. approximately 1.44 times larger than the previous one. Around the approximating value, $\tilde{G}_{\max}$, additional gain values are selected within the range

$$\left[\tilde{G}_{\max} - 3.2 \, , \, \tilde{G}_{\max} + 6.4 \right] \text{dB}. \tag{7.5}$$

For each of these gain values the amplitudes and locations of the pulses are estimated and quantized. Finally, the combination of the quantized parameters that yields the minimum of (7.4) is selected. The quantized value of the best gain approximation from (7.5) and the optimal pulse locations and amplitudes (signs) are transmitted. Combinatorial coding (Cover 1973) is used to transmit the pulse locations.

For rate 5.3 kb/s in order to find the best innovative excitation $v(n)$ a 17-bit algebraic codebook is used. The excitation vector contains, at most, four nonzero pulses. The allowed amplitudes and positions for each of four pulses are given in Table 7.3.

The positions of all pulses can be simultaneously shifted by one (to occupy odd positions) which requires one extra bit. It is seen from Table 7.3 that the last position of each of the last two pulses is outside the subframe boundary, which indicates that the pulse is not present. Each pulse position is encoded by 3 bits and each pulse amplitude

Table 7.4 Bit allocation of the 6.4 kb/s coding algorithm

Parameters coded	Subframe				Total
	0	1	2	3	
LSP indices					24
Adaptive codebook lags	7	2	7	2	18
All the gains combined	12	12	12	12	48
Pulse positions	20	18	20	18	76
Pulse signs	6	5	6	5	22
Grid index	1	1	1	1	4
Total:					192

Table 7.5 Bit allocation of the 5.27 kb/s coding algorithm

Parameters coded	Subframe				Total
	0	1	2	3	
LSP indices					24
Adaptive codebook lags	7	2	7	2	18
All the gains combined	12	12	12	12	48
Pulse positions	12	12	12	12	48
Pulse signs	4	4	4	4	16
Grid index	1	1	1	1	4
Total:					158

is encoded by 1 bit. This gives a total of 16 bits for the four pulses. Furthermore, an extra bit is used to encode the shift resulting in a 17-bit codebook.

The codebook is searched by minimizing the squared error

$$\sum_{n=0}^{59} \left(r(n) - G \sum_{j=0}^{n} h(j)v(n-j) \right)^2$$

between the residual signal and the filtered weighted codeword from the algebraic codebook, where G denotes the codebook gain.

Tables 7.4 and 7.5 present the bit allocations for both high and low bit-rates. These tables mainly differ in the pulse positions and amplitudes coding. Also, at the lower rate only 170 codebook entries are used to quantize the gain vector a of the pitch predictor.

The G.729 Conjugate Structure ACELP (CS-ACELP) was designed for low-delay wireless and multimedia network applications. The input speech sampled at 8 kHz is split into frames 10 ms long. Linear prediction uses a lookahead of a duration equal to 5 ms. This yields a total 15 ms delay. The LP filter excitation is approximated by codewords from two codebooks. The encoder operates at 8 kb/s. The G.729 comes in two versions – the original version is more complex than G.723.1, while the Annex A version is less complex than G.723.1.

The G.728 coder is an example of the backward adaptive LPC-AS coder. It is a hybrid between the lower bit-rate linear predictive analysis-by-synthesis coders (G.723.1 and G.729) and the backward ADPCM coders. The G.728 is a Low-Delay CELP (LD-CELP). Its frame size is equal to five samples. It uses LP analysis to create three different linear prediction filters. The first is a 50th-order prediction filter for the next sample values. The second is a 10th-order prediction filter for the gain values. The third is a perceptual weighting filter that is used to select the excitation signal. There are 1024 possible excitation vectors. They are decomposed into four possible gains (which quantize the difference between the original gain and its predicted value), two possible signs (+ or −), and 128 possible shape vectors. The excitation vector is selected from the given set of vectors by an exhaustive search. Each possible excitation signal is evaluated to find the one that minimizes the MSE in a perceptually weighted space. The perceptual weighting filter defines this space.

The LD-CELP uses a backward adaptive filter that is updated every 2.5 ms. The filter coefficients are not transmitted and the corresponding filter at the decoding side is constructed using the reconstructed (synthesized) speech signal. This is possible since the speech frame is rather short and the synthesized signal is close enough in value to the original signal; therefore, the reconstructed samples can be used for an LP analysis. It is easy to compute that for each five samples the coder transmits gain, sign, and shape by (2+1+7) bits; that is, the bit-rate is equal to $8000/5 \times 10 = 16$ kb/s.

The GSM 06.10 speech codec is used for mobile communications. Its Regular Pulse Excited Linear Predictive (RPELP) coder operates at 13 kb/s. The input speech sampled at 8 kHz is split into frames of 160 samples each; that is, 20 ms long. For each frame, eight LPCs are found by short-term prediction analysis. They are converted into LSPs which are quantized by 36 bits. Each frame is split into four subframes of 5 ms each. For each subframe the lag and the gain of the long-term predictable (pitch) part is found. These estimates are represented by 2 bits and 7 bits, respectively. Together, for each frame, four estimates of lag and gain require $(2 + 7) \times 4 = 36$ bits. The residual after short- and long-term filtering is quantized for each subframe as follows. The 40-sample residual signal is decimated into three possible excitation sequences, each 13 samples long. The sequence with the highest energy is chosen as the best representation of the excitation sequence. The chosen excitation is approximated by a *regular pulse sequence*; that is, a pitch-spaced pulse sequence. Each pulse in this sequence has an amplitude quantized with 3 bits. The position (phase) of the chosen decimated group of excitation samples and the maximum amplitude of the pulse are transmitted by 2 and 6 bits, respectively. In total, four subframes require 188 bits for representation. Thus, each frame is represented by 260 bits and it yields the bit-rate $260/20$ ms $= 13$ kb/s.

There exists also a half-rate version of the GSM codec. It is a so-called Vector Sum Excited Linear Predictive (VSELP) codec operating at a bit-rate of 5.6 kb/s. VSELP uses a linear combination of vectors from two codebooks in order to represent an LP filter excitation.

7.4   LPC vocoders. MELP standard

Typically the LPC vocoder exploits two voicing states: voiced and unvoiced. They determine whether pulse or noise excitation is used to synthesize the output speech. In other words, vocoders are based on the speech production model but do not use the analysis-by-synthesis approach. Vocoders based on classical linear prediction provide highly intelligible speech at extremely low bit-rates but their main shortcoming compared to the LPC-AS codecs is a so-called "machine-like" synthesized speech.

In 1997 the DoD (USA) standardized a 2.4 kb/s MELP vocoder. It improves the quality of the synthesized speech while preserving a rather low bit-rate. The additional features of MELP compared to classical vocoders are:

- converting LPCs to LSPs and using LSPs to represent the LP filter;
- splitting the voice band into five subbands for noise/pulse mixture control and using a mixed pulse-noise excitation;
- shaping pulse excitation;
- using aperiodic pulses for the transition regions between voiced and unvoiced speech segments.

The input speech signal is sampled at 8 kHz and each sample represents a 16 bit integer value. The encoder splits the input sequence into frames of size 180 samples (22.5 ms). Both current and previous frames are used for the analysis. The linear prediction is based on the last 100 samples from the past frame and the first 100 samples in the current frame. A 10th-order LP filter with the transfer function $H_e(z)$ is constructed for each frame by using the autocorrelation method and the Levinson–Durbin recursive procedure. The obtained 10 LPCs are then converted to LSPs.

The pitch detection in the MELP is performed by using autocorrelation analysis. The autocorrelation values $R(p)$ are computed as

$$R(p) = \frac{c_p(0, p)}{\sqrt{c_p(0, 0)c_p(p, p)}}$$

where

$$c_p(i, j) = \sum_{n=\lfloor -p/2 \rfloor -80}^{\lfloor p/2 \rfloor +80} s(n + i)s(n + j), 0 \le p \le 179$$

and p denotes an estimate of the pitch value. Notice that the pitch is found initially on the input speech and later it is refined on the residual signal. The initial pitch estimate is an integer value determined in the range $40 \le p \le 160$.

The determination of voicing in the MELP is performed for five subbands: 0–500 Hz, 500–1000 Hz, 1000–2000 Hz, 2000–3000 Hz, and 3000–4000 Hz. For this purpose the input speech signal is filtered by five sixth-order Butterworth band-pass filters. Then the encoder determines the five voicing strengths vbp_i, $i = 1, 2, \ldots, 5$. It also refines the initial pitch estimate p and the corresponding value $R(p)$. A refined pitch value is found using the first subband signal. A fractional pitch offset Δ is calculated as

$$\Delta = \frac{c_p(0,\, p+1)c_p(p,\, p) - c_p(0,\, p)c_p(p,\, p+1)}{\delta_1 c_p(0,\, p+1) + \delta_2 c_p(0,\, p)}$$

where

$$\delta_1 = c_p(p,\, p) - c_p(p,\, p+1)$$

and

$$\delta_2 = c_p(p+1,\, p+1) - c_p(p,\, p+1).$$

The fractional pitch $p + \Delta$ lies between 20 and 160. The corresponding normalized autocorrelation at the fractional pitch value is

$$R(p + \Delta) = \frac{(1 - \Delta)c_p(0,\, p) + \Delta c_p(0,\, p+1)}{K_\Delta}$$

where

$$K_\Delta = (K_1 + K_2 + K_3)^{1/2}$$
$$K_1 = c_p(0,\, 0)(1 - \Delta)^2 c_p(p,\, p)$$
$$K_2 = 2\Delta(1 - \Delta)c_p(p,\, p+1)$$
$$K_3 = \Delta^2 c_p(p+1,\, p+1).$$

The maximum autocorrelation value $R(p + \Delta)$ in the first subband determines the pitch estimate and is used as voicing strength for the first subband. If $R(p + \Delta) \leq 0.5$, then the aperiodic flag is set to 1; otherwise it is set to 0. If the aperiodic flag is 1, then the decoder uses aperiodic pulse excitation; otherwise the pulse excitation is periodic with the pitch period.

The residual signal is calculated by filtering the input speech by the found LP filter with the transfer function $H_e(z)$. The voicing flags in the other four subbands are calculated separately using the autocorrelation analysis and a peak search in the residual signal about the pitch estimate found in the first subband. The peakiness measure is computed over each 160 samples as

$$\frac{\sqrt{160 \sum_{n=1}^{160} r^2(n)}}{\sum_{n=1}^{160} |r(n)|} \tag{7.6}$$

where $r(n)$ is the nth sample of the residual signal. High peakiness even if the autocorrelation value is low shows that the speech is weakly periodic. If the value (7.6) exceeds 1.34, vbp_1 is set to 1, if it exceeds 1.6, then vbp_i, $i = 1, 2, 3$, are set to 1; that is, the corresponding subbands are classified as voiced subbands.

In order to improve the shape of the pulse in the pulse excitation a 512-point FFT is performed on 200 samples of Hamming windowed residual signal for analysis. The amplitudes of the first 10 harmonics with positions $512\, i/\hat{p}$, $i = 1, 2, \ldots, 10$, are selected where $\hat{p}$ is the reconstructed pitch value. A spectral peak-searching algorithm finds the maximum within $\lfloor 512/\hat{p} \rfloor$ frequency samples around the initial estimate of the position for each of the 10 selected harmonics. The amplitudes of the harmonics

Table 7.6 Bit allocation of the MELP coder

Parameters coded	Voiced	Unvoiced
LSPs	25	25
Fourier amplitudes	8	
Two gains	8	8
Pitch	7	7
Band-pass voicing	4	
Aperiodic flag	1	4
Error protection		13
Sync bit	1	1
Total:	54	58

are vector quantized using a codebook of size 256. In the decoder these quantized amplitudes are used as follows. The inverse DFT (IDFT) is performed on one pitch length $p = \hat{p}$. Since only 10 amplitudes are transmitted but p amplitudes are required for the IDFT, the remaining amplitudes are set to 1. The single pulse excitation is then generated as

$$e_\mathrm{p}(n) = p^{-1} \sum_{k=0}^{p-1} F(k)e^{j2\pi kn/p}$$

where $F(k)$ are the Fourier amplitudes. The phases are set to 0 and since $F(k)$ are real $F(p-k) = F(k)$, $k = 1, 2, \ldots, p/2$. The noise excitation in the decoder is generated by a random number generator. The pulse and noise excitations are then filtered and summed to form the mixed excitation. The pulse excitation filter is given by the sum of all band-pass filter coefficients for the voiced frequency bands. It has the following transfer function:

$$H_\mathrm{p}(z) = \sum_{i=1}^{5} \sum_{k=0}^{p} v_i b_{i,k} z^{-k}$$

where $b_{i,k}$ denotes the filter coefficients and v_i are quantized voicing strengths. The noise excitation is filtered by the filter with the transfer function

$$H_\mathrm{n}(z) = \sum_{i=1}^{5} \sum_{k=0}^{p} (1 - v_i) b_{i,k} z^{-k};$$

that is, this filter is given by the sum of the band-pass filter coefficients for the unvoiced bands. The bit allocation for the MELP coder is presented in Table 7.6. The 10 LSPs are quantized by using a MultiStage Vector Quantizer (MSVQ). The codebook contains four subsets of codewords of size 128, 64, 64, and 64, respectively. The weighted squared Euclidean distance is used as the distortion measure. The approximating vector represents a sum of vectors with one vector chosen from each of the codebooks. If $vbp_1 > 0.6$, then the remaining band-pass voicings are quantized to 1 if their values exceed 0.6 and quantized to 0 otherwise.

7.5 An algorithm for computing linear spectral parameters

Assume that the inverse of the LP filter is described by its transfer function $H_a(z)$,

$$H_a(z) = \frac{1}{1 - a_1 z^{-1} - a_2 z^2 - \cdots - a_m z^{-m}}.$$

Consider the polynomial $A(z)$ which has the form

$$A(z) = A(m)z^{-m} + A(m-1)z^{-(m-1)} + \cdots + A(1)z^{-1} + 1$$

where $A(m) = -a_m$, $A(m-1) = -a_{m-1}, \ldots, A(1) = -a_1/a_m$.

The polynomial $A(z)$ of degree m has exactly m zeros. If the filter is stable, then the zeros of $A(z)$ can be represented as the points located inside the unit circle in the complex plane. Since the coefficients $\{a_i\}$, $i = 1, 2, \ldots, m$, are real numbers, then the zeros of $A(z)$ are complex conjugate numbers. First, using $A(z)$ we construct an auxiliary symmetric polynomial $P(z)$ and an auxiliary antisymmetric polynomial $Q(z)$, both of degree $m + 1$. They have zeros located on the unit circle. Each of the polynomials has a trivial real zero ($z^{-1} = -1$ and $z^{-1} = 1$, respectively). These zeros can be easily removed by dividing by ($z^{-1} + 1$) and ($z^{-1} - 1$), respectively. Therefore, we obtain two polynomials of degree m. Since their zeros are points on the unit circle, they can be given by their phases. Moreover, the zeros are complex conjugate numbers and they have opposite phases. By taking this into account we can reduce each of the two equations of the power m to the equation of the power $m/2$ with respect to the cosines of phases. Then we compute arccosines of the zeros and obtain exactly m numbers φ_i, $i = 1, \ldots, m$, lying in the interval $[0, \pi]$. These numbers determine the frequencies corresponding to the poles of the filter transfer function $H_a(z)$. The lower bound of the range (0) corresponds to the constant component of the signal and the upper bound (π) corresponds to the maximal digital frequency $f_s/2$, where f_s is the sampling frequency. The normalized values $\Phi_i = \varphi_i f_s/(2\pi)$, $i = 1, \ldots, m$, are called *linear spectral parameters*. The detailed description of the algorithm is given below.

Step 1 Construct auxiliary polynomials $P(z)$ and $Q(z)$ using the polynomial $A(z)$. Construct polynomial $P(z)$ of the degree $m + 1$ according to the rule:

$$p_0 = 1, \quad p_{m+1} = 1$$
$$p_k = \begin{cases} A(k) + A(m+1-k), & \text{if } k = 1, 2, \ldots, m/2 \\ p_{m+1-k}, & \text{if } k = m/2 + 1, \ldots, m. \end{cases}$$

For example, for the filter of the second order $m = 2$, we have

$$A(z) = A(2)z^{-2} + A(1)z^{-1} + 1,$$

then

$$P(z) = 1 + (A(2) + A(1))z^{-1} + (A(2) + A(1))z^{-2} + z^{-3}.$$

Construct the polynomial $Q(z)$ of the degree $m + 1$ according to the rule:

$$q_0 = 1, \quad q_{m+1} = -1$$

$$q_k = \begin{cases} A(k) - A(m + 1 - k), & \text{if } k = 1, 2, \ldots, m/2 \\ -q_{m+1-k}, & \text{if } k = m/2 + 1, \ldots, m. \end{cases}$$

For example, for the filter of order $m = 2$, we have

$$Q(z) = 1 + (A(1) - A(2))z^{-1} - (A(1) - A(2))z^{-2} - z^{-3}.$$

It is easy to see that $A(z) = (P(z) + Q(z))/2$.

Step 2 Reducing the degree of the polynomials by 1. Construct a polynomial $PL(z)$ of degree m from the polynomial $P(z)$ as follows:

$$pl_0 = 1$$

$$pl_k = \begin{cases} p_k - pl_{k-1}, & \text{if } k = 1, 2, \ldots, m/2 \\ pl_{m-k}, & \text{if } k = m/2 + 1, \ldots, m. \end{cases}$$

Notice that if the filter is stable, then the obtained polynomial has only complex zeros and they are located on the unit circle.

For example, for the filter of order $m = 2$

$$PL(z) = 1 + (A(1) + A(2) - 1)z^{-1} + z^{-2}$$

that is equivalent to dividing $P(z)$ by $(z^{-1} + 1)$.

Construct $QL(z)$ of degree m as:

$$ql_0 = 1$$

$$ql_k = \begin{cases} q_k + ql_{k-1}, & \text{if } k = 1, 2, \ldots, m/2 \\ ql_{m-k}, & \text{if } k = m/2 + 1, \ldots, m. \end{cases}$$

If the filter is stable, the zeros of $QL(z)$ are complex and are located on the unit circle. For the filter of order $m = 2$

$$QL(z) = 1 + (A(1) - A(2) + 1)z^{-1} + z^{-2};$$

that is, equivalent to dividing $Q(z)$ by $(z^{-1} - 1)$.

Step 3 Reducing the degree to $m/2$. Taking into account that the zeros of $PL(z)$ and $QL(z)$ are located on the unit circle; that is, have the form $z^{-1} = e^{\pm j\omega_i}$, where $i = 1, 2, \ldots, m/2$, we can easily reduce the degree of the equations necessary to solve to find zeros of $QL(z)$ and $PL(z)$. Instead of solving $QL(z) = 0$ and $PL(z) = 0$, we construct polynomials $P^*(z)$ and $Q^*(z)$ of degree $m/2$ according to the formulas:

$$p_0^* = 2^{-m/2}(pl_{m/2} - 2pl_{m/2-2} + 2pl_{m/2-4} - \cdots)$$
$$p_1^* = 2^{-(m/2-1)}(pl_{m/2-1} - 3pl_{m/2-3} + 5pl_{m/2-5} - \cdots)$$
$$p_2^* = 2^{-(m/2-2)}(pl_{m/2-2} - 4pl_{m/2-4} + 9pl_{m/2-6} - \cdots)$$

$$\cdots\cdots\cdots\cdots\cdots\cdots\cdots\cdots\cdots\cdots\cdots\cdots\cdots\cdots\cdots\cdots\cdots$$

$$p_{m/2-1}^* = pl_1/2$$
$$p_{m/2}^* = 1.$$

The formal rule is the following. In the formula for p_0^* in the brackets coefficient 1 is followed by the sequence of alternate coefficients -2 and $+2$. In the remaining formulas the first coefficient is always equal to 1. The signs of the other coefficients alternate. The absolute value of the ith coefficient, $i > 1$, in the series for p_j^* is equal to the sum of absolute values of the $(i - 1)$th coefficient in the series for p_j^* and the ith coefficient in the series for p_{j-1}^*. For example, in the formula for p_2^* we obtain $4 = 3 + 1, 9 = 4 + 5$, then will follow $16 = 7 + 9$ and so on.

The coefficients of $Q^*(z)$ can be obtained analogously from the coefficients of $QL(z)$. In particular for the filter of order $m = 2$ we obtain

$$P^*(z) = (A(2) + A(1) - 1)/2 + z^{-1}$$
$$Q^*(z) = (A(1) - A(2) + 1)/2 + z^{-1}.$$

Step 4 Solve the equations $Q^*(z) = 0$ and $P^*(z) = 0$. These equations have $m/2$ real roots in the form $z_i^{-1} = \cos\omega_i$, $i = 1, 2, \ldots, m/2$. For the filter of order $m = 2$ we obtain

$$z_p^{-1} = -(A(2) + A(1) - 1)/2$$
$$z_q^{-1} = -(A(1) - A(2) + 1)/2.$$

In the general case, for solving equations $Q^*(z) = 0$ and $P^*(z) = 0$ we can use any numerical method of solving algebraic equations. A commonly used method is the following. We search for the roots in two steps. First, we search for approximate root values $root(j)$, $j = 1, 2, \ldots, m/2$, in the closed interval $[-1, 1]$. To do this we split the interval into n equal subintervals, where the value of n depends on the order of the filter. For example, for $m = 10$ we choose $n = 128$. The presence of the root in the subinterval is detected by changing the sign of the polynomial. At the second step we refine the root value. The refined root value $exroot(j)$, $j = 1, 2, \ldots, m/2$, is computed by the formula:

$$exroot(j) = root(j) + 2F(i - 1)/(n(F(i - 1) - F(i)))$$

where i is the number of the subinterval where $root(j)$ is located; $root(j) = -1 + 2(i - 1)/n$; $F(i)$ is the value of the polynomial at the end of the ith subinterval.

Step 5 Denote by z_{pi}^{-1}, $i = 1, 2, \ldots, m/2$, the zeros of $P^*(z)$ and by z_{qi}^{-1}, $i = 1, 2, \ldots, m/2$, the zeros of $Q^*(z)$. Find the LSPs by the formulas

$$\omega_{pi} = \arccos(z_{pi}^{-1}), i = 1, 2, \ldots, m/2$$
$$\omega_{qi} = \arccos(z_{qi}^{-1}), i = 1, 2, \ldots, m/2.$$

Sort the found values in increasing order and normalize them by multiplying by the sampling frequency and dividing by 2π.

The obtained LSPs can be scalar or vector quantized and coded together with other parameters of the speech signal.

The decoder reconstructs the filter coefficients from the quantized LSPs.

Step 1 The quantized LSP values are split into two subsets. The first contains the LSPs with even numbers and the second contains the LSPs with odd numbers. They correspond to the polynomials $P^*(z)$ and $Q^*(z)$, respectively. The LSP values are multiplied by 2π and divided by the sampling frequency. We compute the cosines of the obtained values which represent the zeros of $P^*(z)$ and $Q^*(z)$.

Step 2 Using the Viett theorem we can express the coefficients of $P^*(z)$ and $Q^*(z)$ via their zeros. For the filter of order $m = 2$, we obtain $p_0^* = -z_p^{-1}, q_0^* = -z_q^{-1}, p_1^* = 1, q_1^* = 1$.

Step 3 Using the formulas given above which connect the coefficients of $P^*(z)$ and $PL(z)$ $(Q^*(z), QL(z))$, reconstruct $PL(z), QL(z)$. For the filter of order $m = 2$, we obtain:

$$pl_0 = 1, \; pl_1 = -2z_p^{-1}, \quad pl_2 = 1$$
$$ql_0 = 1, \; ql_1 = -2z_q^{-1}, \quad ql_2 = 1.$$

Step 4 Reconstruct the polynomials $P(z), Q(z)$ of degree $m + 1$ according to the rule:

$$p_0 = 1$$
$$p_j = \begin{cases} pl_j + pl_{j-1}, & \text{if } j = 1, 2, \ldots, m/2 \\ p_{m+1-j}, & \text{if } j = m/2 + 1, \ldots, m + 1 \end{cases}$$

$$q_0 = 1$$
$$q_j = \begin{cases} ql_j - ql_{j-1}, & \text{if } j = 1, 2, \ldots, m/2 \\ -q_{m+1-j}, & \text{if } j = m/2 + 1, \ldots, m + 1. \end{cases}$$

For the filter of order $m = 2$, we obtain

$$p_0 = 1, \quad p_1 = -2z_p^{-1} + 1 = A(2) + A(1), \quad p_2 = p_1, p_3 = 1,$$
$$q_0 = 1, \quad q_1 = -2z_q^{-1} - 1 = A(1) - A(2), \quad q_2 = A(2) - A(1), q_3 = -1.$$

Reconstruct the following polynomials

$$P(z) = 1 + (A(1) + A(2))z^{-1} + (A(1) + A(2))z^{-2} + z^{-3}$$
$$Q(z) = 1 + (A(1) - A(2))z^{-1} + (A(2) - A(1))z^{-2} - z^{-3}.$$

Step 5 Reconstruct $A(z)$:

$$A(z) = (P(z) + Q(z))/2.$$

It is evident that for $m = 2$ this formula gives the correct solution.

Problems

7.1 Quantize the value 125 using the μ-law approximation in Fig. 7.4.

7.2 Assume that the prediction filter is described by the recurrent equation:

$$x_p(n) = \sum_{i=1}^{k} a_i(n-1)x_r(n-i) + e(n)$$

$$e(n) = \sum_{i=1}^{m} b_i(n-1)d(n-i)$$

$$x_r(n) = x_p(n) + d(n).$$

Write down the transfer function of the prediction filter.

7.3 Let the transfer function of a speech synthesis filter have the form $H_a(z) = 1/A(z)$, where

$$A(z) = A(4)z^{-4} + A(3)z^{-3} + A(2)z^{-2} + A(1)z^{-1} + 1$$

$A(1) = -0.7088$, $A(2) = 0.0853$, $A(3) = 0.0843$, $A(4) = 0.1212$. Find the linear spectral parameters of the synthesis filter.

7.4 Reconstruct the prediction filter coefficients using its LSPs. Use the LSP values found by solving Problem 7.3.

7.5 For the given speech file in *.wav format:
Split the speech signal into frames of the given size (N).
For each frame, find the coefficients of the Yule–Walker equations of the given order using the autocorrelation method.
Find the solution of the Yule–Walker equations using the Levinson–Durbin procedure.
Find the ideal excitation (prediction error signal.)

7.6 Perform uniform scalar quantization of the filter coefficients and the error signal obtained by solving Problem 7.5. Compute the compression ratio.
Reconstruct the speech signal using the obtained quantized data. Compute the relative squared approximation error (the ratio of the average squared approximation error to the average energy of the speech signal).

7.7 Find the LSPs of the filter from Problem 7.5. Perform uniform scalar quantization of the LSPs and the error signal. Compute the compression ratio. Reconstruct the speech signal using the obtained quantized data. Compute the relative squared approximation error.

7.8 For each frame of the speech signal from Problem 7.5, plot the amplitude function of the prediction filter. For each frame, plot the amplitude function of the inverse filter.

7.9   Find the pitch period for the given sequence of speech samples:

177	−22	137	378	381	247	−16	−18	108	485
584	372	43	−280	−280	−237	−218	−395	−671	−743
−445	−183	−1	66	−10	−15	74	251	275	144
−177	−390	−176	325	699	755	535	233	123	244
368	335	−83	−495	−508	−235	209	369	149	−211
−284	−9	191	105	−218	−377	−171	63	116	−16
−248	−427	−334	−85	167	190	−17	−154	−18	316
663	675	408	187	−6	16	231	287	93	−168
−463	−514	−292	−2	116	152	0	−42	117	101
−4	−378	−634	−624	−248	139	184	201	115	51
409	697	559	187	−184	−376	−264	50	230	170
−48	−214	−78	188	250	200	20	−261	−501	−557

8 Image coding standards

We use the term *image* for still pictures and *video* for motion pictures or sequences of images. Image coding includes coding a wide range of still images such as bi-level or fax images, monochrome and color photographs, document images containing text, handwriting, graphics, etc.

Similarly to speech signals, images can be considered as outputs of sources with memory. In order to represent an image efficiently, it is important to take into account and remove any observable memory or redundancy. The typical forms of signal redundancy in most image and video signals are *spatial redundancy* and *temporal redundancy*, respectively.

Spatial redundancy takes a variety of different forms in an image. For example, it includes strongly correlated repeated patterns in the background of the image and correlated repeated base shapes, colors, and patterns across an image.

Temporal redundancy arises from repeated objects in consecutive images (frames) of a video sequence. Objects can remain or move in different directions. They can also fade in and out and even disappear from the image.

A variety of techniques used in the modern image- and video-coding standards to compensate spatial redundancy of images is based on transform coding considered in Chapter 5. The second basic principle of image coding is to exploit a human's ability to pay essentially no attention to various types of image distortion. By understanding the masking properties of the human visual system, it is possible to make distortions perceptually invisible.

The main approach to compensate temporal redundancy of video sequences is called *motion compensation* and can be interpreted as a prediction of a current video frame (a group of pixels of the current frame) by the displaced previous frame (a group of pixels of the previous frame). The displacement takes into account that objects' translation movement occurred in the current frame. In this chapter we consider approaches used to extract spatial redundancy from still images. Lossy image compression standards such as JPEG and JPEG-2000 are discussed. Although this book is focused mainly on lossy compression techniques, lossless image compression standards are included for completeness. Chapter 9 will be devoted to methods for compensating temporal redundancy in video sequences.

8.1   Coding of bi-level fax images: JBIG Standard

In this section we consider standards of bi-level image coding. The distinguishing feature of all these standards is that they provide for lossless image coding. The main attribute of these standards is a *compression ratio*, which is the ratio of the original file size in bits divided by the size of the compressed file in bits. We start with the low complexity G3 and G4 coders and then describe basic principles of the Joint Bi-level Image Group (JBIG) standard coder.

Each pixel of bi-level or fax images can be white or black. These two pixel values can be represented by only one bit: 0 corresponds to a black pixel and 1 corresponds to a white pixel. In each bi-level image there are large areas of the same color (black or white). For example, Fig. 8.1 illustrates a line of the bi-level image.

It is easy to see that six successive pixels of the image have the same color (black). We call such a group of pixels a *run* of six pixels each of which has the value 0. The binary representation of the given line is 100000010100.

The G3 fax standard uses a one-dimensional (1D) *run-length coding* of pixels on each line followed by Huffman coding. For our example, the given sequence

$$100000010100$$

is transformed into the following sequence of pairs:

$$(1,1)\ (6,0)\ (1,1)\ (1,0)\ (1,1)\ (2,0)$$

where the first element of the pair is the length of a run and the second element specifies the value of pixels in the run. This sequence of pairs is then coded by the two-dimensional (2D) Huffman code for pairs (*run length, value*). Different Huffman codes are used for coding runs of black and white pixels since their statistics are quite different. For all images the G3 standard uses the same Huffman code specified by a lookup table. To provide error detection, the ends of the lines are indicated by the special code symbol EOL. The G3 standard achieves rather high compression ratios on bi-level text documents.

The G4 fax standard provides an improvement over the G3 fax standard by using a 2D run-length coding which takes into account not only the horizontal spatial redundancy but also the vertical spatial redundancy. In particular, the G4 algorithm uses pixels of the previous line as predicted values for the pixels of the current line. According to the G4 standard, input data are encoded as follows. Starting with the second line, the sum of the previous and the current lines modulo 2 is computed. The obtained sum is processed by a run-length coder followed by a Huffman coder. The end of each line is marked by EOL. The G4 standard coder increases, on the average, the compression ratios on text documents compared to the G3 standard coder.

Figure 8.1   Example of a bi-level image line

The G3, G4 demonstrate rather good performances for text-based documents but they do not provide good compression ratios for documents with handwritten text or gray-scale images converted into binary form. Consequently, a new set of fax standards was created, including the JBIG-1 and JBIG-2 standards.

The key idea behind the JBIG-1 coding is that binary halftone (gray-scale) images have statistical properties that are completely different from binary text and therefore need a different coding algorithm to compress them efficiently. Although halftone images can be converted into binary form using different methods, they have common features, namely, they typically contain rather short runs and need significantly larger than a one line support region for prediction. Gray-scale images can be converted into dot patterns (as in newspapers) or they can be represented by *bit-planes*. Each bit-plane consists of the same order bits of image pixels. Bi-level coding algorithms are applied to each bit-plane separately.

The JBIG-1 standard provides compression ratios which are comparable to the G4 fax standard for text sequences, and significantly improve the compression ratios achieved for binary halftone images.

The JBIG-1 algorithm is complex compared to the G3/G4 standards and we describe here only the basics of this algorithm. The JBIG-1 standard has so-called *sequential* and *progressive* modes. In sequential mode full resolution images are encoded; we do not consider this mode here.

Progressive coding starts with the highest resolution image and ends with the lowest resolution image. The encoder provides a low-resolution base image and a sequence of *delta* files each of which corresponds to another level of resolution enhancement. Each delta file contains the information needed in order to double both the vertical and the horizontal resolution. It allows the user to obtain rather fast a low-resolution image which is less sharp than the original but does not contain any specific distortions. Then, if necessary, the resolution of the received preview can be improved. Progressive mode can also be used for encoding data of different priority. In this case delta files can have lower priority than the lowest resolution image. If for some reason delta files are lost, the user will receive a less sharp image but no entire image areas will be corrupted.

All resolution layers but the lowest are called *differential layers*. A JBIG-1 encoder represents a chain of D differential-layer encoders followed by a bottom-layer encoder. In Fig. 8.2 a block diagram of a JBIG-1 encoder is shown. The image I_j, $j = D$, $D - 1, ..., 1$, at resolution layer j is processed by the jth differential encoder which creates the image I_{j-1} of lower-resolution level and compressed data C_j corresponding to layer j. Block diagrams of the differential-layer encoder and the bottom-layer encoder are shown in Figs 8.3 and 8.4, respectively.

The *resolution reduction* block in Fig. 8.3 downsamples the higher-resolution image I_j, generating lower-resolution image I_{j-1}. Each pixel in the lower-resolution image I_{j-1} replaces four pixels in the higher-resolution image I_j. However, in order to generate a lower resolution image pixel, not four but six of its neighboring higher-resolution pixels and three already processed neighboring lower-resolution image pixels are used. Pixels which determine the value of a lower-resolution image pixel are shown in Fig. 8.5.

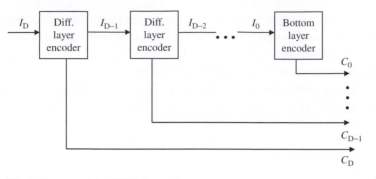

Figure 8.2   Block diagram of the JBIG-1 encoder

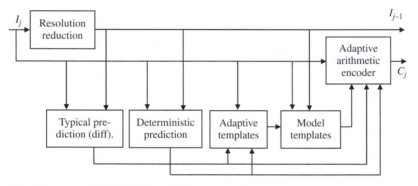

Figure 8.3   Block diagram of JBIG-1 differential-layer encoder

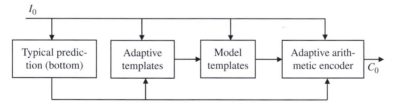

Figure 8.4   Block diagram of the JBIG-1 bottom-layer encoder

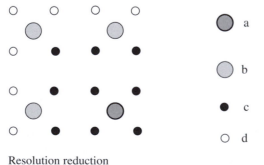

Figure 8.5   Resolution reduction

The circles of "a" and "b" types represent the lower-resolution layer pixels and the circles of "c" and "d" types represent the higher-resolution layer pixels. The value of the "a" type pixel in Fig. 8.5 is determined by the value of a linear combination of six "c" type pixels and three "b" type pixels. If the obtained value exceeds a given threshold, then the "a" type pixel is set to one; otherwise it is set to zero.

The other blocks in Figs 8.3 and 8.4 implement the compression algorithm for image I_j. The most important part in both the bottom-layer coder and the differential-layer coder is an *adaptive arithmetic coder*. This encoder dynamically estimates the probabilities of each pixel to be black (zero) or white (one) depending on the combination of already coded neighboring pixels which form a *pixel context*. More details about context coding can be found in the Appendix. JBIG-1 uses a binary arithmetic coder based on fixed precision integer arithmetics and implemented without multiplications and divisions. Such an arithmetic encoder is called a QM-coder. In fact, the QM-coder combines probability estimating and simplified arithmetic coding. For efficient implementation of the QM-coder, it is important to know which symbol, highly probable or lowly probable, enters the encoder at a given time instant. First, the QM-coder classifies the input bit as More Probable Symbol (MPS) and Less Probable Symbol (LPS). For each input symbol, the QM-coder uses the context to predict which one of the bits (0 or 1) will be MPS. If the predicted MPS bit does not match with the input bit, then the QM-coder classifies such a bit as LPS, otherwise as MPS. An estimate of the conditional probability p of the LPS given the pixel context is then assigned to the LPS. Theoretical and implementation aspects of arithmetic coding are considered in the Appendix. The implementation issues of the QM-coder are given there as well.

The *model-templates block* provides the arithmetic coder with a context value. In other words, it generates an integer S which is determined by the values of neighboring pixels in a pixel context. Pixels whose values determine the context value are called *template* or *model template*. An example of a model template for a differential layer is shown in Fig. 8.6.

For differential-layer coding, a value of S for the "b" type higher-resolution layer pixel is determined by the values of the six previously transmitted pixels (five of them are adjacent "c" type pixels and one is the so-called adaptive "e" type pixel) in the

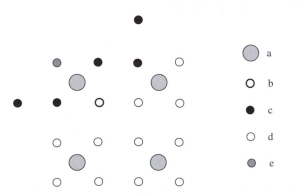

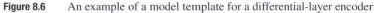

Figure 8.6 An example of a model template for a differential-layer encoder

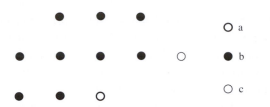

Figure 8.7   An example of a model template for the bottom-layer encoder

higher-resolution image; also by the values of the four "a" type pixels in the already transmitted lower-resolution image, and also by the *spatial phase* of the "b" type pixel to be encoded. The adaptive pixel is spatially separated from the others and can change its position during the encoding. The term "spatial phase" denotes which of the four possible positions the higher-resolution pixel has with respect to its four corresponding lower-resolution pixels. Since each pixel of the pixel context is a single bit, it is easy to compute that there are $4 \times 2^6 \times 2^4 = 4096$ possible context values for the differential-layer coding. An example of a model template for the bottom-layer encoder is shown in Fig. 8.7.

The coding template for the "a" type pixel in the bottom-layer encoder includes 10 already transmitted same resolution pixels: nine adjacent "b" type pixels and one adaptive "c" type pixel which can change its position during the encoding. In total, 1024 different context values are used for bottom-layer coding. An *adaptive templates block* adaptively determines the position of the adaptive pixel with respect to the symbol to be encoded.

In the JBIG standard the probability estimation is implemented as a prediction. A finite-state machine with 112 states is used for the adaptation process. Each state corresponds to each probability estimate. Two arrays $Index(\cdot)$ and $MPS(\cdot)$ of size 4096 each contain the indices (state numbers) of the p values and the MPS values (0 or 1), respectively. At the start, both arrays are initialized to zero. During coding of a symbol having context value S, the LPS probability and MPS value are $p(Index(S))$ and $MPS(S)$, respectively.

The *differential-layer typical prediction* block looks for regions of solid color and represents them by special flags. This block not only increases efficiency of coding but also speeds up the implementation, since none of the processing is performed for pixels belonging to solid color areas.

The *bottom-layer typical predictor* tries similarly to the differential-layer typical prediction to exploit solid color regions of the image to save processing efforts. However, the algorithm differs from that used by the differential-layer typical predictor. This block looks for identical lines of the low-resolution image. If a line is identical to the line above it, it is classified as *typical*. The encoder transmits the flag "typical" for each "typical" line and skips the coding of all pixels in such lines. When the decoder receives the flag "typical," it generates the corresponding line by repetition.

The *deterministic prediction* block uses information from the lower-resolution pixels to exactly determine the values of the next higher-resolution pixels. Deterministic prediction is based on the inversion of the resolution reduction method. For example, assume that the resolution reduction finds each lower-resolution pixel by applying log-ical "AND" to the four associated higher-resolution pixels. Then, if a lower-resolution pixel is equal to one, we can predict that the four associated higher-resolution pixels are also equal to one. The deterministic prediction block flags such pixels and they are not arithmetically encoded. Deterministic prediction is matched with the resolution reduction algorithm and has to be changed if another resolution reduction method is used.

8.2 Coding of halftone images: JPEG Standard

Most halftone image compression standards are considered as lossy compression stan-dards, although they also contain lossless coding modes. Coding efficiency is usually evaluated as a compression ratio. The quality of the synthesized image is characterized by the Signal-to-Noise Ratio (SNR) at the output of the decoder

$$SNR = 10\log_{10}(E_{\text{inp}}/E_{\text{n}})(\text{dB})$$

where E_{inp} denotes the energy of the original signal. For an image of size $M \times N$ it is computed as

$$E_{\text{inp}} = \sum_{i=1}^{M}\sum_{j=1}^{N} p_{i,j}^2$$

where $p_{i,j}$ denotes a pixel value of the image component (Y, U, or V). The value E_{n} is the energy of the quantization noise or, in other words, E_{n} represents the energy of the difference between the original and the reconstructed images. For the image of size $M \times N$ it is computed as

$$E_{\text{n}} = \sum_{i=1}^{M}\sum_{j=1}^{N}(p_{i,j} - \hat{p}_{i,j})^2$$

where $\hat{p}_{i,j}$ is the pixel value of the reconstructed component.

More often, to characterize the synthesized image quality, the Peak Signal-to-Noise ratio (PSNR) is used. It is defined as follows

$$PSNR = 10\log_{10}((255)^2/E_{\text{na}}),$$

where 255 is the maximal pixel value, E_{na} is the average energy of the quantization noise. For the image of size $M \times N$, E_{na} is computed as

$$E_{\text{na}} = \frac{1}{MN}\sum_{i=1}^{M}\sum_{j=1}^{N}(p_{i,j} - \hat{p}_{i,j})^2.$$

The general coder and decoder schemes used in standards for halftone image com-pression are given in Figs 8.8(a), (b). It is easy to see that these schemes describe the

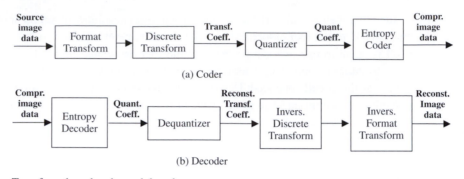

Figure 8.8   Transform-based coder and decoder

transform-based coder and decoder. The existing standards differ mainly in transform types and entropy coders.

JPEG is an acronym for "Joint Photographic Experts Group." It is an international standard for color image compression (JPEG 1994) including two modes: lossless and lossy compression. The JPEG compression algorithm is applied to color images represented in the so-called *RGB* format. This format is a particular case of the more general *Bitmap* (BMP) format which is commonly used for representing halftone and color images in Windows 95, Windows 98, Windows 2000, Windows NT, and Windows XP operation systems. The BMP format of images can represent a noncompressed image as well as a lossless compressed image. Images in the BMP format can be color (4, 8, 16, 24, or 32 bits/pixel) or monochrome (1 bit/pixel).

An image file given in the RGB format contains a header of length 54 bytes followed by R (red), G (green), and B (blue) components of the image. This format implies that the image is noncompressed. Each component is an array of size $M \times N$ bytes for an image of size $M \times N$ pixels. Each byte determines the saturation of the corresponding color. For example, a color image of size 120×160 pixels represented in the RGB format requires $54 + 120 \times 160 \times 3 = 57\,654$ bytes for storing.

The components R, G, and B of an image are highly correlated and are equally important in the sense of image reconstruction. If we apply the same compression algorithm to each of the components, we should not prefer one of them. On the other hand, there exist other color formats called *luminance-chrominance* representations. The luminance describes the intensity of light or *brightness* of the image. It provides a gray-scale version of the image. Two chrominance components carry additional information which converts the gray-scale image into a color image. These two components are related to the *hue* and *saturation*. Hue describes which color is present (red, green, yellow, etc.) in the image. The saturation describes how vivid is the color (very strong, pastel, nearly white). Luminance-chrominance representations are particularly important for good image compression, since their components are almost uncorrelated. Moreover, the most important information is concentrated in the luminance component. Thus, we do not lose much information if we decimate chrominance components. One of the luminance-chrominance representations is called the YUV format and another is called the YCbCr format.

Since YUV and YCbCr formats are more convenient from the image compression point of view, usually color image compression coders convert the RGB format to the YUV or YCbCr format. It can be done by using the following linear transforms

$$Y = 0.299R + 0.587G + 0.114B$$

$$U = (B - Y)0.5643$$

$$V = (R - Y)0.7132$$

and

$$Y = 0.299R + 0.587G + 0.114B$$

$$Cb = (B - Y)0.5643 + 128$$

$$Cr = (R - Y)0.7132 + 128.$$

The inverse transforms can be described as follows

$$G = Y - 0.714V - 0.334U$$

$$R = Y + 1.402V$$

$$B = Y + 1.772U$$

and

$$G = Y - 0.714(Cr - 128) - 0.334(Cb - 128)$$

$$R = Y + 1.402(Cr - 128)$$

$$B = Y + 1.772(Cb - 128).$$

Then, the U and V components are decimated by a factor of 2. In other words, four neighboring pixels which form the square of size 2×2 are described by the four values of component Y, one value of component U, and one value of component V. Each chrominance value is computed as the rounded-off arithmetic mean of the corresponding four pixel values belonging to the considered square. As a result we obtain the so-called YUV 4:1:1 standard video format which is typically used as the input format for most image- and videocodecs. It is easy to compute that in using this format we spend only 6 bytes for each square of size 2×2 instead of 12 bytes spent by the original YUV format. Thus, due only to decimation of the components U and V, we already compress the image by a factor of 2 without any visible distortions.

The JPEG standard is based on the 2D-DCT-II. The component Y and decimated components U and V are processed by blocks of size 8×8 pixels. Since the 2D DCT for blocks 8×8 is a separable transform, it can be performed by first applying a 1D transform to the rows of each 8×8 block and then applying a 1D transform to the columns of the resulting block, that is,

$$Y_{k,l} = \frac{c_k}{2} \sum_{i=0}^{7} \left[\frac{c_l}{2} \sum_{j=0}^{7} X_{i,j} \cos \left(\frac{(2j+1)l\pi}{16} \right) \right] \cos \left(\frac{(2i+1)k\pi}{16} \right)$$

where

$$c_k = \begin{cases} 1/\sqrt{2}, & \text{if } k = 0 \\ 1, & \text{if } k \neq 0 \end{cases}$$

and $X_{i,j}$ denotes the pixel of the image component (Y, U, or V) and $Y_{k,l}$ is the transform coefficient.

The inverse transform can be written

$$X_{i,j} = \sum_{k=0}^{7} \frac{c_k}{2} \left[\sum_{l=0}^{7} \frac{c_l}{2} Y_{k,l} \cos \left(\frac{(2j+1)l\pi}{16} \right) \right] \cos \left(\frac{(2i+1)k\pi}{16} \right).$$

Performing a 2D-DCT is equivalent to decomposing the original image block using a set of 64 2D cosine basis functions. These functions are created by multiplying a horizontally oriented set of eight 1D basis functions (cosine waveforms sampled at eight points) by a vertically oriented set of the same functions. By convention, the coefficient $Y_{0,0}$ that scales the product of two constant 1D basis functions is in the upper left corner of the array of 64 transform coefficients. This coefficient is called the DC coefficient, whereas the rest of the coefficients are called AC coefficients. The names DC (direct current) and AC (alternating current) appeared from the historical use of the DCT for analyzing electrical currents. The frequencies of the horizontal basis functions increase from left to right and the frequencies of the vertical basis functions increase from top to bottom.

Example 8.1   In Fig. 8.9 an 8×8 block of gray-scale image is shown. The matrix X containing pixel values of this 8×8 block has the form

$$X = \begin{pmatrix} 47 & 52 & 53 & 49 & 76 & 171 & 201 & 199 \\ 50 & 51 & 56 & 53 & 88 & 179 & 202 & 202 \\ 50 & 54 & 55 & 58 & 91 & 173 & 205 & 205 \\ 53 & 59 & 61 & 66 & 92 & 156 & 196 & 200 \\ 54 & 55 & 69 & 81 & 102 & 150 & 176 & 181 \\ 48 & 60 & 83 & 90 & 113 & 154 & 175 & 173 \\ 49 & 78 & 96 & 98 & 121 & 160 & 179 & 177 \\ 62 & 94 & 115 & 111 & 130 & 167 & 178 & 175 \end{pmatrix}.$$

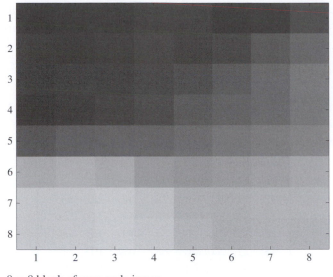

Figure 8.9 8×8 block of gray-scale image

The corresponding matrix Y of the DCT coefficients is

$$Y = \begin{pmatrix} 907.28 & -416.90 & 90.49 & 18.43 & -71.09 & 11.31 & 15.25 & -8.62 \\ -42.53 & -67.13 & 68.73 & 27.93 & -13.67 & 8.20 & 8.37 & -9.23 \\ 24.55 & 8.72 & -10.39 & -1.25 & -16.65 & -5.53 & 5.02 & -2.22 \\ -24.60 & 4.63 & -7.93 & -1.39 & 5.71 & 3.56 & 1.82 & 0.66 \\ 1.36 & 11.60 & 5.38 & -4.75 & 2.28 & 1.61 & -2.21 & -1.43 \\ -2.06 & -6.70 & 7.42 & -4.08 & -4.01 & -1.03 & -5.17 & 0.31 \\ 0.56 & -0.41 & -1.41 & -0.01 & -0.51 & 1.77 & 0.25 & 0.16 \\ -2.14 & -0.53 & -0.13 & 0.76 & 1.43 & -0.14 & -0.61 & -1.32 \end{pmatrix}. \quad (8.1)$$

It is easy to see that DCT localizes the signal energy in a few coefficients located in the left upper corner of the transformed block.

The obtained transform coefficients are quantized by a uniform scalar quantizer. The generalized Gaussian distribution with parameter $\alpha \leq 0.5$, considered in Chapter 3, provides a useful model for the 1D probability distribution of DCT-II coefficients. As follows from Fig. 3.8 the lattice vector quantizer applied to the output of such a source wins at most 0.1 bit per sample compared to the scalar quantization with extended zero zone. However, scalar quantization with extended zero zone is noticeably more efficient than the uniform scalar quantization for this source. This explains why scalar quantization is commonly used in nowadays image- and videocodecs but, on the other hand, in more modern codecs uniform scalar quantization is almost completely replaced by scalar quantization with extended zero zone.

The quantization is implemented as rounding-off the DCT coefficients divided by a quantization step. The values of the steps are set individually for each DCT coefficient, using criteria based on the visibility of the basis functions. Thus, the quantized coefficient is

$$Z_{k,l} = \left[Y_{k,l}/q_{k,l} \right] = \left\lfloor (Y_{k,l} \pm q_{k,l}/2)/q_{k,l} \right\rfloor, k, l = 0, 1, \ldots, 7$$

where $q_{k,l}$ is the klth entry of the quantization matrix Q of size 8×8. The standard JPEG uses two different quantization matrices. The first matrix is used to quantize the luminance component (Y) and has the form

$$Q = \begin{pmatrix} 16 & 11 & 10 & 16 & 24 & 40 & 51 & 61 \\ 12 & 12 & 14 & 19 & 26 & 58 & 60 & 55 \\ 14 & 13 & 16 & 24 & 40 & 57 & 69 & 56 \\ 14 & 17 & 22 & 29 & 51 & 87 & 80 & 62 \\ 18 & 22 & 37 & 56 & 68 & 109 & 103 & 77 \\ 24 & 35 & 55 & 64 & 81 & 104 & 113 & 92 \\ 49 & 64 & 78 & 87 & 103 & 121 & 120 & 101 \\ 72 & 92 & 95 & 98 & 112 & 100 & 103 & 99 \end{pmatrix}.$$

The second matrix is used to quantize the chrominance components (U, V) and is given by

$$Q = \begin{pmatrix} 17 & 18 & 24 & 47 & 99 & 99 & 99 & 99 \\ 18 & 21 & 26 & 66 & 99 & 99 & 99 & 99 \\ 24 & 26 & 56 & 99 & 99 & 99 & 99 & 99 \\ 47 & 66 & 99 & 99 & 99 & 99 & 99 & 99 \\ 99 & 99 & 99 & 99 & 99 & 99 & 99 & 99 \\ 99 & 99 & 99 & 99 & 99 & 99 & 99 & 99 \\ 99 & 99 & 99 & 99 & 99 & 99 & 99 & 99 \\ 99 & 99 & 99 & 99 & 99 & 99 & 99 & 99 \end{pmatrix}.$$

In order to obtain larger compression ratios, all entries of the quantization matrices except the value $Q(0, 0)$ can be multiplied by a proper scaling coefficient. Notice that in the recent implementations of the JPEG standards the DC coefficient is typically quantized with step equal to 8. All AC coefficients are quantized with equal quantization step which determines the quality level of the reconstructed image.

Example 8.2   Continuing Example 8.1, assume that the AC coefficients from 8.1 are quantized with step equal to 12; then the matrix of quantized coefficients has the form

$$Z = \begin{pmatrix} 113 & -35 & 8 & 2 & -6 & 1 & 1 & -1 \\ -4 & -6 & 6 & 2 & -1 & 1 & 1 & -1 \\ 2 & 1 & -1 & 0 & -1 & 0 & 0 & 0 \\ -2 & 0 & -1 & 0 & 0 & 0 & 0 & 0 \\ 0 & 1 & 0 & 0 & 0 & 0 & 0 & 0 \\ 0 & -1 & 1 & 0 & 0 & 0 & 0 & 0 \\ 0 & 0 & 0 & 0 & 0 & 0 & 0 & 0 \\ 0 & 0 & 0 & 0 & 0 & 0 & 0 & 0 \end{pmatrix}.$$

The matrix Z contains a small number of significant coefficients and a large number of zeros.

The quantized DCT coefficients are coded by a variable-length coder. Since the DC coefficient of a block is the measure of the average value of 64 block pixels, it is not surprising that the DC coefficients of neighboring blocks are usually highly correlated. The AC coefficients are almost uncorrelated, which explains why the DC coefficient is encoded separately from the AC coefficients. The coding procedure is performed in two steps. At the first step the DC coefficient is DPCM encoded by subtracting from it the DC coefficient value of the previous already encoded 8×8 block and the AC coefficients are coded by a run-length coder. At the second step the obtained values are coded by a Huffman code.

Assume that DC_i and DC_{i-1} denote the DC coefficients of the ith and $(i-1)$th blocks, respectively. If DC_{i-1} is already encoded, then in order to encode the coefficient DC_i we compute the difference $DC_i - DC_{i-1}$ and encode it. For gray-scale images (or one of the Y, U, V components of a color image), a pixel is represented by 8 bits. It is easy to verify that the DC coefficient belongs to the range $0 \leq Y_{0,0} \leq 255 \times 64/8$. Thus, the difference $DC_i - DC_{i-1}$ takes on values from the range $[-2047, 2047]$. This range is split into 12 categories where the ith category includes the differences with the length of their binary representation equal to i bits. These categories are the first 12 categories shown in Table 8.1. Each DC coefficient is described by a pair *(category, amplitude)*. If the value $DC_i - DC_{i-1} \geq 0$, then the *amplitude* is the binary representation of this value of length equal to the *category*. If $DC_i - DC_{i-1} < 0$, then the *amplitude* is the codeword of the complement binary code for the absolute value of $DC_i - DC_{i-1}$ which also has length equal to *category*. The category value is then coded by the Huffman code.

Example 8.3 Let $DC_{i-1} = 191$ and $DC_i = 180$. Then the difference is $DC_i - DC_{i-1} = -11$. It follows from Table 8.1 that the value -11 belongs to the category 4. The binary representation of value 11 is 1011 and the codeword of the complementary code is 0100. Thus, the value -11 is represented as (4,0100). If the codeword of the Huffman code for 4 is, for example, 110, then -11 is coded by the codeword 1100100

Table 8.1 Categories of integer numbers

Category	Numbers
0	0
1	-1,1
2	-3,-2,2,3
3	-7,...,-4,4,...,7
4	-15,...,-8,8,...,15
5	-31,...,-16,16,...,31
6	-63,...,-32,32,...,63
7	-127,...,-64,64,...,127
8	-255,...,-128,128,...,255
9	-511,...,-256,256,...,511
10	-1023,...,-512,512,...,1023
11	-2047,...,-1024,1024,...,2047
12	-4095,...,-2048,2048,...,4095
13	-8191,...,-4096,4096,...,8191
14	-16383,...,-8192,8192,...,16383
15	-32767,...,-16384,16384,...,32767
16	32768

of length 7. The decoder first processes the category value (in our case it is 4); then, the next 4 bits correspond to the value of $DC_i - DC_{i-1}$. Since the most significant bit is equal to 0, the value is negative. Inverting bits, we obtain the binary representation of 11. Notice that using categories simplifies the Huffman code. Without using categories we would need the Huffman code for an alphabet of a much larger size; that is, coding and decoding would be much more complicated.

For gray-scale images or Y, U, or V components, the AC coefficients can take on values from the range $[-1023, 1023]$. After quantization, many of these coefficients become zeros. Thus, it is necessary to code only a small number of nonzero coefficients, simply indicating before their positions. To do this efficiently the 2D array of the DCT coefficients is rearranged into a 1D linear array by scanning in zigzag order as shown in Fig. 8.10. This zigzag index sequence creates a 1D vector of coefficients where the DCT coefficients, which are amplitudes of the basis functions with lower frequencies, tend to be at lower indices. When the coefficients are ordered in such a way, the probability of coefficients being zero is an approximately monotonic increasing function of the index.

The *run-length* coder generates a codeword $((run\text{-}length, category), amplitude)$ where *run-length* is the length of zero run followed by the given nonzero coefficient, *amplitude* is the value of this nonzero coefficient, and *category* is the number of bits needed to represent the *amplitude*. The pair (*run-length*, *category*) is coded by a 2D Huffman code and the *amplitude* is coded as in the case of DC coefficients and is added to the codeword.

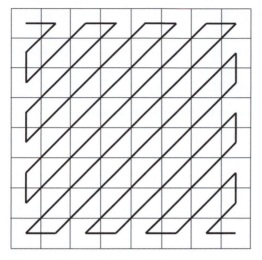

Figure 8.10   Zigzag ordering of DCT coefficients

Example 8.4   Let the nonzero coefficient preceded by six zeros be equal to −18. It follows from Table 8.1 that −18 belongs to category 5. The codeword of the complement code is 01101. Thus, the coefficient is represented by ((6,5), 01101). The pair (6,5) is coded by a Huffman code and the value 01101 is added to the codeword. If the codeword of the Huffman code for (6,5) is, for example, 1101, then the codeword for −18 is 110101101.

There are two special cases when we encode the AC coefficients:

- After a nonzero coefficient all other AC coefficients are zero. In this case the special symbol (EOB) is transmitted which codes the End-Of-Block condition.
- A pair (*run-length, category*) appeared which is not included in the table of the Huffman code. In this case a special codeword called *escape-code* followed by the uniform codes for run-length and nonzero value are transmitted.

Example 8.5   Continuation of Example 8.1. After zigzag scanning of the DCT coefficients for the 8×8 block, we obtain:

$$-35, -4, 2, -6, 8, 2, 6, 1, -2, 0, 0, -1, 2, -6, 1, -1, 0, -1, 1, 0, 0, -1, 0, 0,$$
$$-1, 1, 1, -1, 1, 0, 0, 0, 1, 0, 0, 0, 0, 0, 0, 0, 0, -1. \tag{8.2}$$

The run-length coded sequence (8.2) is:

$$((0, 6), -35), ((0, 3), -4), ((0, 2), 2), ((0, 3), -6), ((0, 4), 8), ((0, 2), 2)),$$

$$((0, 3), 6), ((0, 1), 1), ((0, 2), -2), ((2, 1), -1), ((0, 2), 2), ((0, 3), -6),$$

$$((0, 1), 1), ((0, 1), -1), ((1, 1), -1), ((0, 1), 1), ((2, 1), -1), ((2, 1), -1),$$

$$((0, 1), 1), ((0, 1), 1), ((0, 1), -1), ((0, 1), 1), ((3, 1), 1), ((8, 1), -1).$$

After representing the amplitudes in the binary form we have

$$((0, 6), 011100), ((0, 3), 011), ((0, 2), 10), ((0, 3), 001), ((0, 4), 1000),$$

$$((0, 2), 10), ((0, 3), 110), ((0, 1), 1), ((0, 2), 01), ((2, 1), 0), ((0, 2), 10),$$

$$((0, 3), 001), ((0, 1), 1), ((0, 1), 0), ((1, 1), 0), ((0, 1), 1), ((2, 1), 0), ((2, 1), 0),$$

$$((0, 1), 1), ((0, 1), 1), ((0, 1), 0), ((0, 1), 1), ((3, 1), 1), ((8, 1), 0).$$

The main shortcoming of the JPEG standard is the so-called "blocking artifacts" which appear in the reconstructed image. These specific distortions arise because the transform and quantization are applied to blocks of the image.

8.3   JPEG-LS

The standard JPEG-LS is intended for efficient lossless and near lossless compression of gray-scale and color still images. It was developed later than the DCT-based lossy compression JPEG standard considered in the previous section. JPEG-LS improves the lossless compression mode of the JPEG standard and, unlike the lossy compression mode of JPEG, its near lossless compression mode guarantees that the maximum error between each pixel of the original image and the same pixel of the reconstructed image is less than or equal to a predetermined value. In this section we focus on the lossless compression mode of the JPEG-LS standard.

This mode is based on the so-called LOw COmplexity LOssless COmpression algorithm (LOCO) (Weinberger *et al.* 1996) which contains the following main steps:

- prediction
- context modeling
- entropy coding.

Prediction of the current pixel x is based on the neighboring already processed pixels a, b, c, d as shown in Fig. 8.11.

The predicted value $\hat{x}$ is calculated as follows:

$$\hat{x} = \begin{cases} \min\{a, b\}, & \text{if } c \geq \max\{a, b\} \\ \max\{a, b\}, & \text{if } c \leq \min\{a, b\} \\ a + b - c, & \text{otherwise.} \end{cases} \tag{8.3}$$

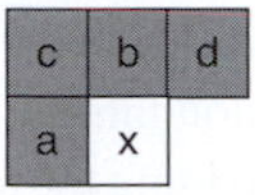

Figure 8.11   Neighboring pixels used for prediction of the current pixel x

The underlying idea of such a prediction is to detect edges in the neighborhood of the current pixel. The three conditions in (8.3) correspond to the presence of horizontal or vertical edges or to their absence. For example, if $b < a$ and $c > a$, then the first condition describes a vertical edge separating the "bright area" (a, c) from the "dark area" (b, x). The first condition corresponds to a transition from the "bright area" to the "dark area" while the second condition describes a transition from the "dark area" to the "bright area." If the third condition is true, then we can conclude that the area surrounding the pixel x is relatively smooth. If there exists a vertical edge to the left of the current pixel, then we choose $\hat{x} = b$. If a horizontal edge is found above the current pixel, then $\hat{x} = a$ is chosen. Otherwise, that is, if no edges are detected, then $\hat{x} = a + b - c$.

The obtained prediction error $e = x - \hat{x}$ is then entropy coded with probabilities determined by context values. The context is based on local gradients, i.e. differences $D_1 = d - b$, $D_2 = b - c$, and $D_3 = c - a$. Gradients reflect a degree of smoothness of the area surrounding the current pixel which determines the statistical behavior of the prediction errors. In order to limit the number of contexts, the gradients D_1, D_2, and D_3 are quantized into nine cells indexed $-4, -3, \ldots, 0, \ldots, 3, 4$. It gives $9 \times 9 \times 9 = 729$ different context values. The number of context values is then reduced to 365 by merging symmetric contexts. Let us consider a triplet (q_1, q_2, q_3), where q_i is a quantized gradient value, $i = 1, 2, 3$. If q_1 is negative, then instead of e we encode $-e$ using the context $(-q_1, -q_2, -q_3)$. It means that we distinguish $4 \times 9 \times 9$ different context values. If $q_1 = 0$ and q_2 is negative, we encode $-e$ instead of e using the context $(0, -q_2, -q_3)$. It gives 4×9 context values additionally. If $q_1 = q_2 = 0$ and q_3 is negative, then we encode $-e$ using $(0, 0, -q_3)$, i.e. five more possible contexts. In total, we obtain $4 \times 9 \times 9 + 4 \times 9 + 5 = 365$ different contexts. In other words, the contexts (q_1, q_2, q_3) and $(-q_1, -q_2, -q_3)$ are merged or the probabilities $P(e|(q_1, q_2, q_3))$ and $P(-e|(-q_1, -q_2, -q_3))$ are equated.

In order to further reduce the number of parameters in the context modeling, it is assumed that the prediction errors from a fixed predictor in gray-scale images can be modeled by a two-sided geometric distribution centered at zero. Such a distribution can be described by only one parameter: the rate of the exponential decay. However, for context-conditioned predictors, this distribution can have a bias. As a result it is described by two parameters: the rate of the exponential decay and the shift from zero. Thus, instead of estimating a conditional probability distribution, only two values are calculated per each context. The first value is the so-called *bias cancellation* C_x which is subtracted from the prediction error e, i.e. $e' = e - C_x$. This can be interpreted as an adaptive part of the prediction, depending on the context value. The calculation of the bias cancellation for each context is based on an accumulated prediction error for this context and a number of context occurrences. However, to lower the computational complexity in fixed precision integer arithmetics, it is implemented by a more sophisticated method than simply dividing the accumulated error by the number of context occurrences. We do not consider this implementation issue.

The second parameter k, implicitly characterizing the exponential decay of the conditional distribution, determines a parameter $T = 2^k$ of the Golomb–Rice codes used for entropy coding. It is known that the Golomb–Rice codes (for details, see the Appendix) are optimal for sources of one-sided geometrically distributed non-negative integer values. To apply them to the stream of corrected (biased) prediction errors, the prediction errors are first mapped into non-negative values according to the formula

$$\hat{e} = \begin{cases} 2e', & \text{if } e' \geq 0 \\ 2|e'| - 1, & \text{otherwise.} \end{cases}$$

According to (A.14) the parameter T of the Golomb–Rice code should be chosen as the mathematical expectation of the one-sided geometric distribution. For each context the mathematical expectation of the corresponding distribution is estimated as the average accumulated prediction error for this context. To avoid the division operation, the parameter T is computed as

- $T = 0$
- while $(N << T) < S$
- $T \leftarrow T + 1$

where S is the accumulated error and N is the number of occurrences of the context.

Using $T = 2^k$ significantly simplifies encoding and decoding. Encoding of a given value x is reduced to appending the k least significant bits to the unary code of $\lfloor x/T \rfloor + 1$. For example, let $T = 4$, $x = 9$, then we obtain the codeword: 00101.

The JPEG-LS standard has two coding modes. One already considered is called *regular mode*. The second is called *run mode* and is used in the case indicated by the all-zero gradients, i.e. $D_1 = D_2 = D_3 = 0$. In this mode the Golomb–Rice code is used for coding run lengths. First, the binary sequence of prediction error positions is coded by the Golomb–Rice code. The run 0^T, i.e. the sequence of $T = 2^k$ zeros is coded by 0. The subsequences $0^l 1$, where $l < T$, are coded as 1 followed by the binary representation of the number l of length k. Then, the amplitudes of the prediction errors are coded as in the regular mode.

8.4   Standard JPEG-2000

The standard JPEG-2000 (JPEG-2000 2000) is based on the same scheme (see Fig. 8.8) as the JPEG standard. Unlike the JPEG standard, it uses a discrete wavelet transform which is applied to the whole image (more precisely to its components). The discrete wavelet transform is followed by quantization and lossless coding (entropy coding).

The generalized Gaussian distribution with parameter $\alpha \leq 0.5$ (see Chapter 3) can be considered as a proper model for the 1D probability distribution of the wavelet transform coefficients. From consideration of Fig. 3.8 we can conclude that the rate-distortion function of scalar quantization with an extended zero zone is rather close to the rate-distortion function of lattice vector quantization for such a source. This explains that two different quantization procedures are allowed by the standard: scalar quantization with an extended zero zone and an optionally used lattice quantization.

The standard JPEG-2000 contains a long list of different features and options. We give only a short overview of the main approaches and techniques described in the standard.

The image components (Y, U, V) are divided into rectangular blocks called *tiles*. The 2D discrete wavelet transform can be applied either to the entire image components (Y, U, V) or, optionally, to their tiles. Since the 2D wavelet transform is a separable transform, it is implemented as two 1D discrete wavelet transforms applied first to all rows of the original matrix and then to all columns of the resulting matrix. In such a way the image components or tiles are decomposed into wavelet subbands. The subband coefficients corresponding to a given step of decomposition form a *resolution layer*.

The subband coefficients are then quantized and collected into rectangular arrays called *precincts*. One quantization step is allowed for each subband. Each precinct is partitioned into *codeblocks*. The codeblocks are processed independently. The coding algorithm is called Embedded Block Coding with Optimized Truncation (EBCOT). Typical codeblocks are of size 64×64.

All coefficients of the codeblocks are represented by their bit-planes, that is, binary arrays of bits of the same order from all coefficients of a codeblock. The codeblocks are processed consecutively, bit-plane by bit-plane, starting with the most significant bit-plane and ending with the least significant bit-plane. Each of the bit-planes is scanned stripe-by-stripe as shown in Fig. 8.12.

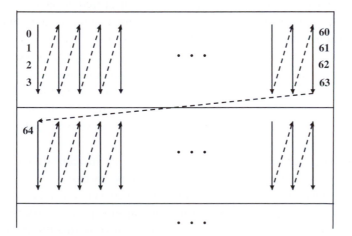

Figure 8.12   Bit-plane scanning order

d_0	v_0	d_1
h_0	x	h_1
d_2	v_1	d_3

Figure 8.13   Context forming for significance propagation pass

The first bit-plane of the codeblock containing a nonzero bit is processed by so-called *cleanup pass*. Each bit of the remaining bit-planes is encoded in one of three passes:

- significance propagation
- magnitude refinement
- cleanup pass.

All three coding passes use the so-called *significance map*. This map connects each coefficient of the codeblock with a binary variable called *significance state*. A coefficient becomes significant when the most significant bit equal to one is found. The significance state of such a coefficient is set to one; otherwise it is equal to zero. For each coding pass different rules for forming bit contexts are used. The context vector consists of significance state values of at most eight neighboring coefficients with respect to the chosen bit (see Fig. 8.13). In Fig. 8.13 x denotes a bit to be encoded and h_i, v_i, and d_i, $i = 0, 1$, are significance state values of horizontal, vertical, and diagonal neighboring coefficients, respectively. All possible context vectors are mapped into 17 allowed context values. The context value together with a given bit value is provided to the QM-coder (a simplified binary context arithmetic code also used by the JBIG standard).

In the *significance propagation pass* only bits of coefficients whose significance states have not yet been set to one and having at least one of its eight neighbors significant are coded. If the encoded bit at the same time becomes equal to one, then its significance state is set to one and the sign of the corresponding coefficient is encoded. Otherwise the significance state remains zero. The significance map of eight neighboring coefficients is mapped into one of nine allowed context values. The mapping of the sum of the significance state values into context values depends on the subband.

Another context vector is used for sign bit coding. The diagonal neighbors are not used in this case. Each neighbor can be significant positive, significant negative, or insignificant. The nine possible horizontal and nine possible vertical configurations are mapped into five context values.

During the *magnitude refinement pass*, bits of the coefficients which are already significant are encoded. Three context values are used in this pass.

During the *cleanup pass*, bits of the coefficients which were previously insignificant and were not coded during the last significance propagation pass are processed. Beside the context shown in Fig. 8.13 a *run-length* context is used during this pass. The run-length context is used if all four neighboring coefficients in a column of the scan (see

Fig. 8.12) are insignificant and each has only insignificant neighbors. One bit is used to specify whether or not the column is all-zero. If all bits in the column are zero, then the symbol zero is sent to the arithmetic coder with the run-length context. Otherwise, the symbol one is sent together with the run-length context and the length of the zero-run (from 0 to 3) represented by two bits.

Using wavelet decomposition and embedded block coding based on bit-plane coding, the JPEG-2000 provides different types of *scalability* of its bitstream. *Spatial scalability* based on wavelet decomposition implies that the bitstream can be decomposed at different resolution levels. The so-called *SNR scalability* based on bit-plane coding implies that the bitstream can be decomposed at different quality levels (SNRs).

Another important property of the JPEG-2000 is the possibility to encode different regions of the image with a different quality. The standard supports *Region Of Interest (ROI)* coding. Specified regions can be encoded with higher quality than other regions of the image. Two methods of ROI coding are included: the *general scaling-based* method and the *MAXSHIFT* method. The idea of these methods is the following. We scale (shift) coefficients in order to move the bits associated with ROI to higher bit-planes. For the *MAXSHIFT* method the scaling value is chosen in such a way that the minimum ROI coefficient is larger than the maximum non-ROI coefficient. It allows us to have ROI of different shapes without storing or transmitting any ROI mask.

In Fig. 8.14 the original size 1000×1000 gray-scale image "Castle" is shown.[1] In Figs 8.15 and 8.16 the same image reconstructed from the 50 times compressed JPEG and JPEG-2000 data, respectively, is shown. It is easy to see that the images processed by the JPEG codec have blocking artifacts near the 8×8 block boundaries while the images processed by the JPEG-2000 codec have specific distortions around sharp edges. Distortions typical for codecs based on the DWT can be seen in the picture fragment presented in Fig. 8.17.

The distinguishing feature of wavelet-based coding is that, even for rather high compression ratios, the main content of the image is preserved. In Figs 8.18 and 8.19 the image "Castle" reconstructed from the 100 times compressed JPEG and JPEG-2000 data, respectively, is shown. From comparison of Figs 8.18 and 8.19, it follows that the quality of wavelet-based coding with compression ratio 100 is superior to the block-based JPEG coding with the same compression ratio. Moreover, the quality of wavelet-based codecs can be further improved by avoiding restrictions related to bit-plane coding (see, for example, Bocharova *et al.* 1997, 1999). The image "Castle" reconstructed from the 100 times compressed data obtained by one of the DWT-based codecs is shown in Fig. 8.20.

Notice that the main shortcoming of wavelet-based coding is that the reconstructed images are blurred even for rather low compression ratios. The quality of JPEG encoded images outperforms the quality of the JPEG-2000 encoded images for compression ratios of the order 10–20.

---

[1] The photo of Lichtenstein Castle was taken by the author, June 2007.

Figure 8.14   Original gray-scale image "Castle"

Figure 8.15   Image "Castle" reconstructed from the 50 times compressed JPEG data
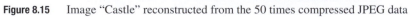

Figure 8.16   Image "Castle" reconstructed from the 50 times compressed JPEG-2000 data

Figure 8.17   Specific distortions of the DWT-based codec

Figure 8.18   Image "Castle" reconstructed from the 100 times compressed JPEG data

Figure 8.19   Image "Castle" reconstructed from the 100 times compressed JPEG-2000 data

Figure 8.20   Image "Castle" reconstructed from the 100 times compressed DWT coefficients

Problems

8.1   Consider the bi-level image described by the matrix

$$
\begin{pmatrix}
0 & 1 & 1 & 1 & 0 & 0 & 0 & 0 & 1 & 0 & 1 & 1 & 0 & 0 & 0 & 0 \\
0 & 1 & 0 & 1 & 1 & 1 & 0 & 0 & 0 & 0 & 0 & 0 & 0 & 0 & 1 & 1 \\
1 & 0 & 1 & 1 & 0 & 0 & 1 & 0 & 1 & 1 & 1 & 1 & 1 & 1 & 1 & 1 \\
1 & 1 & 1 & 1 & 0 & 0 & 0 & 0 & 1 & 0 & 1 & 1 & 1 & 0 & 0 & 0 \\
0 & 0 & 0 & 0 & 0 & 0 & 1 & 0 & 1 & 0 & 1 & 1 & 1 & 0 & 0 & 0 \\
1 & 1 & 0 & 0 & 0 & 1 & 0 & 0 & 0 & 0 & 0 & 0 & 0 & 1 & 0 & 0 \\
0 & 0 & 0 & 1 & 1 & 0 & 0 & 0 & 0 & 0 & 1 & 1 & 1 & 1 & 0 & 0 \\
1 & 1 & 1 & 0 & 0 & 0 & 0 & 0 & 1 & 1 & 1 & 0 & 0 & 1 & 1 & 1 \\
1 & 1 & 1 & 1 & 0 & 0 & 0 & 0 & 1 & 0 & 1 & 1 & 1 & 0 & 0 & 0 \\
0 & 0 & 0 & 0 & 0 & 0 & 1 & 0 & 1 & 0 & 1 & 1 & 1 & 0 & 0 & 0 \\
1 & 1 & 0 & 0 & 0 & 1 & 0 & 0 & 0 & 0 & 0 & 0 & 0 & 1 & 0 & 0 \\
0 & 1 & 1 & 1 & 0 & 0 & 0 & 0 & 1 & 0 & 1 & 1 & 0 & 0 & 0 & 0 \\
0 & 1 & 1 & 1 & 0 & 0 & 0 & 0 & 1 & 0 & 1 & 1 & 0 & 0 & 0 & 0 \\
0 & 0 & 0 & 1 & 1 & 1 & 0 & 0 & 0 & 0 & 0 & 0 & 0 & 1 & 0 & 1 \\
0 & 0 & 0 & 1 & 1 & 1 & 0 & 0 & 0 & 0 & 0 & 0 & 0 & 1 & 0 & 1 \\
1 & 1 & 1 & 1 & 1 & 1 & 0 & 0 & 1 & 1 & 1 & 0 & 1 & 1 & 1 & 1
\end{pmatrix}.
$$

Encode the image using run-length coding according to the G3 standard.

8.2   Encode the bi-level image described in Problem 8.1 according to the G4 standard.

8.3   Let the DC coefficients of two successive blocks be equal to 161 and 178, respectively. Construct the codeword for DC coding.

8.4 Consider the following matrix of quantized DCT coefficients

$$
\begin{pmatrix}
173 & -28 & 23 & -5 & -1 & 0 & 0 & 0 \\
65 & 39 & -11 & 3 & 0 & 1 & 0 & 0 \\
24 & 6 & -4 & 9 & 0 & 0 & -1 & 0 \\
-9 & -2 & 0 & 0 & 0 & 0 & 0 & 0 \\
0 & 0 & 0 & 0 & -3 & 0 & 0 & 0 \\
-2 & 1 & 0 & 0 & 0 & 0 & 0 & 0 \\
0 & 0 & 0 & 0 & 0 & 0 & 0 & 0 \\
0 & 0 & 0 & 0 & 0 & 0 & 0 & 0
\end{pmatrix}.
$$

Construct the codewords for run-length coding.

8.5 Transform an image given in BMP format into YUV format. Decimate the U, V components and compute the PSNR for the reconstructed U and V components. Plot the PSNR as a function of the decimation coefficients. Perform the YUV to RGB transform. Compare the reconstructed and the original images for different decimation coefficients.

8.6 Compute the DCT coefficients for the Y component obtained when solving Problem 8.5. Perform uniform scalar quantization of the obtained coefficients. (Recommendation: quantize the DC coefficient with a step equal to 8 and consider quantization of AC coefficients with steps 2, 4, 8, 16, 32.) Reconstruct the Y component using the IDCT for the quantized coefficients. Plot the PSNR (for Y) as a function of the quantization step.

8.7 Encode the quantized transform coefficients for the Y component obtained when solving Problem 8.6 using a run-length variable encoder. Estimate the entropy of the obtained streams of run lengths and levels. Plot the PSNR (for Y) as a function of the estimated compression ratio.

8.8 Perform a wavelet decomposition of the component Y in Problem 8.5 using the following wavelet filters:

$$h_0(n) = (-1, 2, 6, 2, -1)/\sqrt{32}$$
$$h_1(n) = (-2, 4, -2)/\sqrt{32}$$
$$g_0(n) = (2, 4, 2)/\sqrt{32}$$
$$g_1(n) = (-1, -2, 6, -2, -1)/\sqrt{32}.$$

8.9 Perform uniform scalar quantization of the wavelet coefficients obtained when solving Problem 8.8. (Recommendation: use different quantization steps for different subbands.) Reconstruct the Y component from the quantized wavelet coefficients using inverse wavelet filtering.

8.10 Estimate the entropy of the quantized wavelet coefficients from Problem 8.9 for each subband. Estimate the compression ratio. Plot the PSNR as a function of the estimated compression ratio.

9　Video-coding standards

Video signals represent sequences of images or frames which can be transmitted with a rate from 10 up to 60 frames/s (fps) providing the illusion of motion in the displayed signal. They can be represented in different formats which differ in frame size and number of fps. For example, the video format QCIF intended for video conferencing and mobile applications uses frames of size 176×144 pixels transmitted at rate 10 fps. The High Definition Digital Television (HDTV) standard uses frames of much larger size, 1280×720 pixels, transmitted at rate 60 fps. The frames can be represented in RGB or YUV formats with 24 or fewer (due to decimation of U and V components) number of bits per pixel. If the frames are in RGB format, then the first processing step in any video coder is the RGB to YUV transform.

Unlike images, video sequences contain the so-called temporal redundancy which arises from repeating objects in consecutive frames of a video sequence. The simplest method of temporal prediction can use the previous frame as a prediction of the current frame. However, the residual formed by subtracting the prediction from the current frame typically has large energy, the reason being the object movements between the current and the previous frames. Better predictions can be performed by compensating the motion between two frames.

Different *motion compensation* techniques are used to compensate the temporal redundancy of video sequences. The main idea behind these methods is to predict the displacement of a group of pixels (usually a block of pixels) from their position in the previous frame. Information about this displacement is represented by so-called *motion vectors* which are transmitted together with the DCT coded difference between the predicted and original images.

9.1　Motion compensation method

Since motion compensation is a key element in most video coders, we begin by considering the basic concepts of this processing step. The most widely used method of motion compensation is called *block matching motion compensation*. It is based on compensating movement of blocks of size $M \times N$. A shifted co-sited (i.e. with the same spatial position) block from the previous frame is used as the prediction. The way that the motion estimator works is illustrated in Fig. 9.1. Each $M \times N$ pixel block in the current frame is compared with a set of blocks in the previous frame in order to choose one that

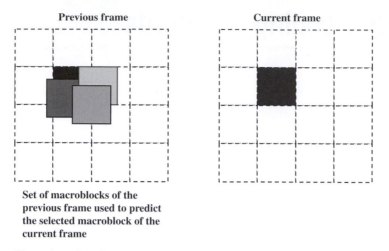

Set of macroblocks of the
previous frame used to predict
the selected macroblock of the
current frame

Figure 9.1   Illustration of motion-compensated coding

represents the best prediction of the current block with respect to a given matching criterion. The set of candidate blocks in the previous frame is constructed as a set of shifts
of the co-sited block in the previous frame. This set of blocks includes blocks within a
limited region of the co-sited block. A popular matching criterion is the energy of the
residual formed by subtracting a block belonging to the set of candidate blocks from the
current $M \times N$ block. In other words, to find the best matching block, we search for

$$\min_{\alpha, \beta} \left\{ \sum_{m=0}^{M-1} \sum_{n=0}^{N-1} \left(x_c(m, n) - x_p(m + \alpha, n + \beta) \right)^2 \right\} \tag{9.1}$$

where $x_c(m, n)$ and $x_p(m, n)$ denote pixels of the current block and of the co-sited
block of the previous frame, respectively, α and β are shifts of the pixel along the
coordinate axes. For blocks of size $M \times N$ (typically, $M = N = 16$) we search for the
best prediction in the range $-M/2 \le \alpha \le M/2, -N/2 \le \beta \le N/2$.

The chosen best matching block becomes the prediction of the current block and
is subtracted from it to obtain a residual block. The coordinates of the displacement
between the current block and the position of the block-prediction form the corresponding coordinates of the *motion* vectors. More precisely, for the ith block of
the current frame we transmit a pair of coordinates (α_i, β_i) which shows how we
should translate the corresponding block of the previous frame. The motion vector
$A = (\alpha_1, \alpha_2, \ldots, \alpha_{N_B})$ contains α-coordinates for all blocks and the motion vector
$B = (\beta_1, \beta_2, \ldots, \beta_{N_B})$ contains β-coordinates for all blocks, where N_B denotes the
number of blocks in the frame. These motion vectors are entropy-coded and transmitted
or stored as a part of the compressed data. The difference between the current and the
motion-compensated frames is transformed using DCT, quantized, entropy-coded, and
transmitted or stored together with coded motion vectors.

Figure 9.2 shows block diagrams of motion-compensated image coder and decoder.
The key idea used in the encoder is to combine transform coding (in the form of DCT

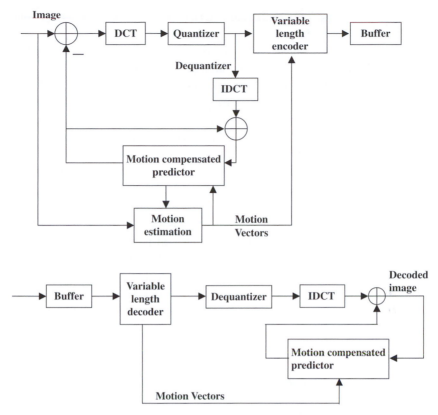

Figure 9.2   Motion-compensated coder and decoder

of 8×8 pixel blocks) with predictive coding (in the form of differential PCM). The first step in the motion-compensated image coder is to create a motion-compensated prediction error using macroblocks of 16×16 pixels. This computation requires only a single frame to be stored in the receiver. Notice that for color images, motion compensation is performed only for the Y component of the image. The decimated motion vectors obtained for Y are then exploited to form motion-compensated U and V components. The resulting error signal for each of Y, U, and V components is transformed using DCT, quantized by an adaptive quantizer, entropy-encoded using a variable-length coder, and buffered for transmission over a fixed-rate channel. The main problem relating to the block matching motion compensation is its high computational complexity. In order to minimize (9.1) for each macroblock we have to perform an exhaustive search over all admissible pairs (α, β). If $-M/2 \le \alpha \le M/2$ and $-N/2 \le \beta \le N/2$, then we search over $(M+1)(N+1)$ shifts of the given macroblock. If $M = N = 16$, it means that we search over $17^2 = 289$ shifts of each macroblock. To speed up the search procedure, different versions of the so-called *logarithmic search* procedure are used.

The logarithmic search can be, for example, organized as follows. Assume that we search in the ranges $-4 \le \alpha \le 4$ and $-4 \le \beta \le 4$. It means that we have to consider $9^2 = 81$ shifts of the co-sited macroblock in the previous frame. Instead of this, we first search over pairs: (0,4), (4,0), (0,0), (4,4), (−4,0), (0,−4), (−4,−4), (4,−4), (−4,4). We obtain the vectors $A_1 = (\alpha_{11}, \alpha_{12}, \ldots, \alpha_{1N_B})$ and $B_1 = (\beta_{11}, \beta_{12}, \ldots, \beta_{1N_B})$. Then the components of the found vectors are used as centers for the next search which is performed over the pairs: (0,2), (2,0), (0,0), (2,2), (−2,0), (0,−2), (−2,−2), (2,−2), (−2,2); that is, for the ith macroblock we search over $(\alpha_{1i}, \beta_{1i} + 2)$, $(\alpha_{1i} + 2, \beta_{1i})$, $(\alpha_{1i}, \beta_{1i})$, $(\alpha_{1i} + 2, \beta_{1i} + 2)$, $(\alpha_{1i} - 2, \beta_{1i})$, $(\alpha_{1i}, \beta_{1i} - 2)$, $(\alpha_{1i} - 2, \beta_{1i} - 2)$, $(\alpha_{1i} + 2, \beta_{1i} - 2)$, $(\alpha_{1i} - 2, \beta_{1i} + 2)$. Now we obtain the vectors $A_2 = (\alpha_{21}, \alpha_{22}, \ldots, \alpha_{2N_B})$ and $B_2 = (\beta_{21}, \beta_{22}, \ldots, \beta_{2N_B})$. Then we use the components of the obtained vectors A_2 and B_2 as centers for the next search which we perform over the pairs: (0,1), (1,0), (0,0), (1,1), (−1,0), (0,−1), (−1,−1), (1,−1), (−1,1). Thus, for the ith macroblock we search over the pairs: $(\alpha_{2i}, \beta_{2i} + 1)$, $(\alpha_{2i} + 1, \beta_{2i})$, $(\alpha_{2i}, \beta_{2i})$, $(\alpha_{2i} + 1, \beta_{2i} + 1)$, $(\alpha_{2i} - 1, \beta_{2i})$, $(\alpha_{2i}, \beta_{2i} - 1)$, $(\alpha_{2i} - 1, \beta_{2i} - 1)$, $(\alpha_{2i} + 1, \beta_{2i} - 1)$, $(\alpha_{2i} - 1, \beta_{2i} + 1)$. At this step we find the final motion compensation vectors A and B which we encode, store or transmit together with the compressed prediction error. It is easy to see that using the above-described procedure, we search over only 27 shifts of the macroblock instead of 81. The price to be paid for this reduction of the computational complexity of search procedure is a loss of optimality. However, this nonoptimal search procedure usually does not lead to a significant loss in the compression ratio for a given quality.

9.2   Overview of video-coding standards

All known video-coding standards can be divided into a few groups:

- video standards for video conferencing. The ITU standards H.261 (H.261 1993) for ISDN video conferencing, H.263 (H.263 2005) standard for POTS video conferencing, and H.262 for broadband video conferencing all belong to this group;
- video standards for storing movies on CD with 1.2 Mb/s allocated to video coding and 256 kb/s allocated to audio coding. For example, ISO-MPEG-1 standard (MPEG-1 1993);
- video standards for storing video on DVD, with 2–15 Mb/s allocated to video and audio coding. For example, ISO-MPEG-2 standard (MPEG-2 1999);
- video standards for low-bit-rate video telephony over POTS networks, with 10 kb/s allocated to video and 5.3 kb/s allocated to voice coding. The ITU H.324 standard and ISO-MPEG-4 Part 2 standard (MPEG-4 2002) belong to this group. Notice that the H.263 standard represents the video part of H.324 and the G.723 speech standard is its part for voice coding;
- the ITU H.264 standard that is equivalent to the ISO-MPEG-4 Part 10 standard (MPEG-4 AVC) (H.264 2007; MPEG4-AVC 2005; Richardson 2003) was developed to provide good video quality of significantly lower bit-rates than the MPEG-2, H.263 or MPEG-4 Part 2 standards. Additionally, the standard is more flexible than previous standards and can be applied to a wide variety of applications;
- video standards for HDTV, with 15–400 Mb/s allocated to video coding.

Although there is a large variety of video standards, almost all of them are based on the same principle; the corresponding coders contain the same blocks and differ mainly from each other in block parameters. All coders first determine the type of frame using some criterion. The so-called *INTRA* or *I*-frame is coded and transmitted as an independent frame. In other words, I-frames are coded as still images. In order to encode the I-frames, coders use a DCT on blocks of size 8×8 pixels and the corresponding part of the coder is the same as that used by the JPEG coder described in Chapter 8 (see Fig. 8.8). An initial frame is always classified as an I-frame. Other I-frames usually correspond to the frames where scenes change.

Frames between two I-frames are supposed to contain the same scene which slowly changes from frame to frame owing to small motions of the objects in the scene. They are coded efficiently in the *INTER* mode using the motion compensation technique. The corresponding part of the coder has the form shown in Fig. 9.2.

9.2.1 H.261 and its derivatives

The H.261 video codec was initially intended for ISDN teleconferencing. It is a basis video mode for most modern multimedia conferencing systems. The H.262 video codec is essentially the high-bit-rate MPEG-2 standard and will be described later in this chapter. The H.263 low-bit-rate video codec is intended for use in POTS teleconferencing at modem rates of 14.4–56 kb/s, where the modem rate includes video coding, speech coding, control information, and other types of datum.

The H.261 standard is intended for conferencing applications with only small, controlled amounts of motion in a scene, and with rather limited views consisting mainly of head-and-shoulders views of people, along with the background. It supports such video formats as the CIF and QCIF formats.

The H.263 video codec is based on the same DCT and motion compensation techniques used in H.261. However, since this standard is designed for conferencing through narrowband POTS, it is improved compared to the H.261 standard. The following enhancements are included:

- Half-pixel motion compensation is used in order to increase the accuracy of estimating best matching blocks. This feature significantly improves the prediction capability of the motion compensation algorithm in cases where there is object motion that needs fine spatial resolution for accurate estimation.
- Improved variable-length coding (for example, arithmetic coding is used instead of the Huffman coding).
- Advanced motion prediction modes including overlapped block motion compensation are exploited.
- A motion prediction mode that combines a bidirectionally predicted image with a forward predicted image is included.

The bidirectional prediction means that there are two types of predicted or INTER frames. The first type is called *P frames*. They are predicted from the most recently reconstructed I or P frame. There are also *bidirectional* or *B frames*. They are predicted

As stored or transmitted

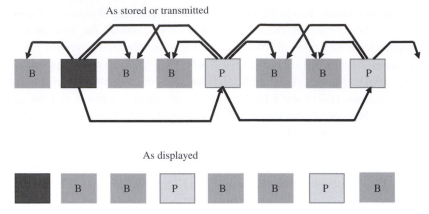

As displayed

Figure 9.3   Sequence of frames with bidirectional prediction

from the closest two I or P frames, one in the past and one in the future. To search for matching blocks in those frames, we first try the forward prediction (using the previous I or P frame), then the backward prediction (using the next I or P frame), and then we average the two blocks from the future and past frames and subtract that average from the block to be encoded. Figure 9.3 illustrates the sequence of frames in the case of bidirectional prediction. In addition, H.263 supports a wider range of image formats including 4CIF (704 × 576 pixels) and 16 CIF (1408 × 1152 pixels).

9.2.2   MPEG-1, MPEG-2, and MPEG-4

This series of standards intended for digital compression of audio and video signals was developed by the Motion Picture Experts Group formed by the ISO (International Organization for Standardization).

The MPEG-1 standard is a multimedia standard with specifications for coding, compression, and transmission of audio, video, and data streams in series of synchronized mixed packets. It is optimized for storage of multimedia content on standard CD-ROM or applications at about 1.5 Mb/s. More precisely, this standard was designed to allow 74 minutes of digital video to be stored on CD. The supported image formats are 352 × 288 pixels at 25 fps and 352 × 240 pixels at 30 fps.

The video coding in MPEG-1 is very similar to the video coding of the H.26X series. The video signal is processed by doing spatial coding using the DCT-II of 8 × 8 pixel blocks, quantizing the DCT coefficients for each block in a zigzag scan, and applying a variable run-length coding to the resulting DCT coefficient stream. Temporal coding is performed by using the ideas of uni- and bidirectional motion-compensated prediction.

MPEG-2 is an extension of the MPEG-1 standard. It is aimed at broadcast formats at higher data rates. It efficiently codes the so-called *interlaced* video and supports a wide range of bit-rates. The MPEG-2 also provides tools for the *scalable* coding where useful video can be reconstructed from pieces of the total bitstream. Similarly to the JPEG-2000 image coding (see Chapter 8) the total bitstream may be structured in layers, starting with a base layer (that can be decoded by itself) and adding refinement layers to reduce quantization distortion or improve resolution. In order to understand specific features of the MPEG-2 standard, we shall consider what is meant by the term *interlaced*.

As mentioned before, a movie is a sequence of images (frames) displayed at a given rate. PAL TV is video displayed at 25 fps and NTSC TV at 30 fps. It is enough to display video frames at 25 or 30 fps to provide for human eye illusion of moving images but on the TV screen the image would be perceived to be flickering. To reduce flicker it is necessary to update the frames with higher rate but this can be done only at the cost of worse resolution, since the television bandwidth is limited. It was found that flicker can be avoided by displaying the same frame in two parts with double rate for each part. These two parts of the image are called *fields*. The first field consists of the odd lines and the second field consists of the even lines of the frame. Displaying the fields with the rate 60 half-fps for NTSC and 50 half-fps for PAL video avoids flicker. Although only half the lines are updated each time, the viewer perceives each frame with full resolution owing to the afterglow of the screen and the persistence of human vision. When a line is drawn on the TV screen, for a short time we continue to see it even after it has actually faded. Therefore, TV is interlaced video: 60 half-images/s for NTSC and 50 half-images/s for PAL. It is important to note that one interlaced frame contains fields from two instants in time. In a video camera the second field, which contains only the even-numbered lines of the frame, is sampled 20 ms after the first field.

In video systems other than television, noninterlaced video is commonly used. In noninterlaced video, all the lines of the frame are sampled at the same instant in time. It is also called *progressively scanned* or *sequentially scanned* video. Computer monitors are almost all progressively scanned monitors. Digital television standards include both interlaced and noninterlaced modes. The H.261, H.263, and MPEG-1 standards work with noninterlaced progressively scanned video.

The MPEG-2 standard allows us to encode standard-definition television video at bit-rates from about 3–15 Mb/s as well as high-definition television video at 15–30 Mb/s. It is used for storing movies on DVD. Notice that MPEG video coding standards are parts of more general standards which include audio coding. Most MPEG-2 decoders can decode MPEG-1 bitstreams.

The MPEG-2 coder uses the same principle as the MPEG-1 coder. Its motion-compensated predictor uses the block matching motion compensation technique but supports a variety of methods to generate a prediction. For example, the block may be *forward predicted* from a previous frame, *backward predicted* from a future frame, or *bidirectionally predicted* by using the average of a forward and a backward prediction. The method used to predict the macroblock can be changed from one macroblock to the other. Additionally, the two fields within a macroblock can be predicted separately with their own motion vectors, or together using common motion vectors. One more option is to make a *zero-value prediction* which implies that the current macroblock is not predicted at all and is itself DCT coded instead of the prediction error. For each macroblock to be coded, the coder chooses between these prediction modes, trying to maximize the decoded frame quality within the constraints on the bit-rate. The choice of prediction mode is transmitted to the decoder as a part of the output bitstream.

The MPEG-2 standard supports a wide range of different applications. To implement all the features of the entire standard in all decoders is inefficient from the point of view of complexity and required bandwidth. In order to restrict the parameters and tools for a variety of practical applications, the standard is divided into subsets called *profiles* and *levels*. A profile is a subset of algorithmic tools such as a compression algorithm. For example, a profile may or may not use scalability. A level defines a set of constraints on parameter values (such as frame size, bit-rate, or the number of layers supported by scalable profiles). A decoder, which uses a given profile and level, supports only the corresponding subset of algorithmic tools and set of parameter constraints.

Two nonscalable profiles are defined by the MPEG-2. The *simple* profile uses no B-frames, and hence no backward or bidirectional prediction. It does not require frame reordering which would add about 120 ms to the coding delay. This profile is suitable for low-delay applications such as video conferencing, where the overall delay is about 100 ms.

The *main* profile is the most widely used profile. It supports B-pictures that increases the synthesized video quality but adds about 120 ms to the coding delay to allow for the frame reordering. Main profile decoders can decode MPEG-1 video as well.

There are several scalable profiles. The *SNR* profile provides SNR or quality scalability. The output bitstream consists of layer bitstreams which can be transmitted sequentially, starting with the base-layer bitstream. The higher priority bitstream contains only base-layer data which correspond to a certain image quality. Bitstream of a lower priority refinement layer can be added to the base-layer bitstream to obtain a higher quality image. Examples of SNR-scalable coder and decoder are shown in Figs 9.4 and 9.5, respectively.

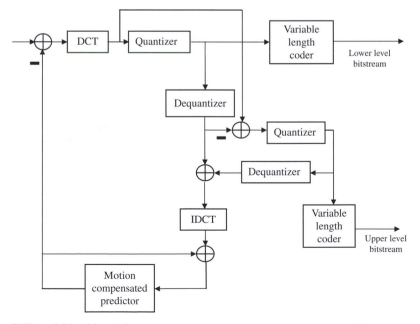

Figure 9.4   SNR-scalable video coder

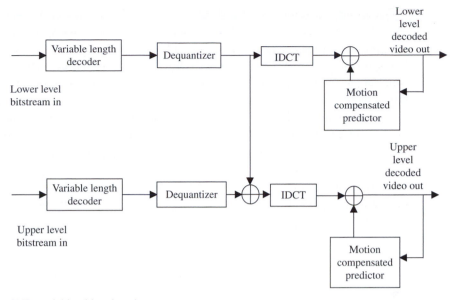

Figure 9.5   SNR-scalable video decoder

The codec operates in a similar manner to the nonscalable codec but it includes an extra quantization step. The coder quantizes the DCT coefficients with a given step, variable-length encodes them, and transmits them as the lower-level or base-layer bitstream. The quantization error introduced by the first quantizer is then itself quantized, variable-length encoded, and transmitted as the upper-level or enhancement-layer bitstream. Motion vectors are transmitted only at the base layer.

The base-layer bitstream can be decoded in the same way as the nonscalable case shown in Fig. 9.2. In order to improve the received image quality, the enhancement-layer coefficient refinements (quantization error of the base layer) are dequantized and added to the base-layer coefficients. The resulting coefficients are then decoded as in the nonscalable case. The SNR-scalable video decoder is shown in Fig. 9.5.

The MPEG-4 standard is not only more efficient than the MPEG-2 standard in the sense of compression video and audio signals but it also significantly improves the previous video standards in terms of flexibility. Its application areas include digital television, streaming video over the Internet, mobile multimedia, digital consumer video cameras, and many others. The concept of profiles and levels used in the MPEG-2 standard was further developed in the MPEG-4 standard. Moreover, it has such new distinguishing features as, for example, video object coding; that is, coding of arbitrary-shaped regions of a video scene. The standard also contains the error resilience tools providing effective transmission over practical networks. The MPEG-4 supports both interlaced and progressively scanned video and is intended for a large range of bit-rates. It is optimized for three bit-rate ranges:

1. below 64 kb/s
2. 64–384 kb/s
3. 384 kb/s–4 Mb/s.

One of the key features of the MPEG-4 is *object-based coding*. A visual scene can consist of one or more objects. A video object is a region of the video scene that can have an arbitrary shape and can exist an arbitrarily long time. An object in a given frame is called a Visual Object Plane (VOP). It can be natural or synthetic and in the simplest case it can be an entire rectangular frame. Each object is characterized by a shape, motion, and texture.

The binary matrix representing the shape of a VOP is called a *binary mask*. In this mask every pixel belonging to the VOP is set to one, and all other pixels are set to zero. This mask is then partitioned into Binary Alpha Blocks (BABs) of size 16×16. There are three types of BAB: internal BABs which lie entirely inside the VOP, external BABs which do not belong to the VOP, and boundary BABs which only partially belong to the VOP. The *gray-scale mask* of the VOP is a matrix where each entry is either an 8-bit integer, if the corresponding pixel belongs to the VOP, or zero. The gray-scale mask is also split into 16×16 *alpha blocks* corresponding to BABs. Each *alpha block* is encoded separately. To encode *alpha blocks* more efficiently, those macroblocks which have pixels of the same value are flagged and coded in a special way. If all pixels of the macroblock are zero, it is called a *transparent* macroblock. For a transparent macroblock, no shape coding is necessary. Neither is texture information coded for such macroblocks. If all pixels of the macroblock belong to the VOP, then the macroblock is called an *opaque* macroblock. For the macroblocks flagged as opaque, the shape coding is not necessary but the texture needs to be coded. To encode binary shape information for the boundary macroblocks, the corresponding BABs are coded by using context arithmetic coding. The gray-scale shape information is coded by using motion compensation and DCT-based encoding.

The approaches for motion compensation in the MPEG-4 standard are similar to those used in other coding standards. The main difference is that the block-based techniques used in the other standards are adapted to the VOP structure used in MPEG-4. MPEG-4 provides three modes for encoding an input VOP:

- A VOP may be encoded independently of any other VOP. In this case the encoded VOP is called an *INTRA VOP* (I-VOP).
- A VOP may be predicted (using motion compensation) based on another previously decoded VOP. Such VOPs are called *Predicted VOPs* (P-VOPs).
- A VOP may be predicted based on past as well as future VOPs. Such VOPs are called *Bidirectional Interpolated VOPs* (B-VOPs). B-VOPs may only be interpolated based on I-VOPs or P-VOPs.

For internal macroblocks, motion compensation is performed in the usual way, based on block matching of 16×16 macroblocks as well as 8×8 blocks (in advanced prediction mode). This results in two motion vectors for the entire macroblock, and two for each of its blocks (in advanced prediction mode). Motion compensation is performed with half-pixel precision.

For macroblocks that only partially belong to the VOP, motion vectors are estimated using the *modified block (polygon) matching* technique. As a criterion of matching, this method uses the Sum of the Absolute Differences (SAD) computed for only those pixels in the macroblock that belong to the VOP. In the case the reference macroblock in the previous frame lies on the VOP boundary, a *repetitive padding* technique assigns values to pixels outside the VOP. The SAD is then computed using these padded pixels also.

MPEG-4 also supports overlapped motion compensation which usually results in better prediction at lower bit-rates. For each block of the macroblock, the motion vectors of neighboring blocks are considered. The resulting motion vector for each pixel is formed as a weighted average of all motion vectors.

In the case of an I-VOP, the texture information is given by the corresponding pixels of the luminance Y and chrominance U and V components. In the case of motion-compensated VOPs, the texture information represents the prediction error obtained after motion compensation. For encoding the texture information, the standard 8×8 block-based DCT is used. To encode an arbitrary-shaped VOP, an 8×8 grid is superimposed on the VOP. Using this grid, 8×8 internal macroblocks are encoded without modifications. Boundary macroblocks contain arbitrarily shaped texture data. A padding technique is used to extend these shapes into rectangular macroblocks. The luminance component is padded on a 16×16 basis, while the chrominance blocks are padded on an 8×8 basis. Repetitive padding consists in assigning a value to the pixels of the macroblock that lie outside of the VOP. When the texture data are the prediction error after motion compensation, the blocks are padded with zero-values. Padded macroblocks are then coded using the standard DCT-based technique.

One of the functionalities supported by MPEG-4 is the mapping of static textures into 2D or 3D surfaces. MPEG-4 has a separate mode for encoding static texture information. The static texture-coding technique is based on wavelet transforms. This technique provides a high degree of scalability, more than the DCT-based texture-coding technique. The wavelet coefficients are quantized and encoded using a zerotree algorithm and arithmetic coding.

A *sprite* coding is a part of an object-based coding strategy. The idea of this method is to decompose all frames in a video sequence into static background and moving foreground objects. Background sprite consisting of all pixels belonging to the background is essentially a static image visible throughout a video sequence. It can move or warp in certain limited ways when the camera zooms and rotates over the scene. The background sprite can be transmitted only once, at the beginning of the transmission. The encoder estimates and transmits the camera parameters, indicating how the sprite should be moved or warped in order to create the background of the next frames. However, how to generate a sprite image from a raw video sequence automatically is still an open issue, because the motion of foreground video objects not only disturbs the accuracy of the motion estimation but also blurs the generated sprite image. Sprite-based coding is very suitable for synthetic objects.

Similar to the representation of VOPs, the texture information for a sprite is represented by one luminance component and two chrominance components. The three components are processed separately, but the methods used for processing the chrominance components are after appropriate scaling, the same as those used for the luminance component. Shape and texture information for a sprite is encoded as for an I-VOP.

9.2.3   Standard H.264

This standard is named H.264 by ITU-T and MPEG-4 part 10 or MPEG-4/AVC by ISO/IEC. It is the newest video-coding standard intended for various applications such as video broadcasting, video streaming, video conferencing, and HDTV. The H.264 has higher efficiency and better subjective quality than previous ISO standards (MPEG-1, MPEG-2, MPEG-4 part 2) and previous ITU standards (H.261, H263). We will not give a complete description of this complex standard here. Our goal is to show the distinguishing features of the H.264 standard that improve performances, compared to previous standards.

One of the important features of the H.264 standard is combining transform coding with *intra-prediction in spatial domain*. A prediction of the current block is based on the pixels from neighboring already encoded and reconstructed blocks. The difference between the predicted and the original block is then coded by using a transform coding technique. Nine intra-prediction modes for 4×4 blocks and four intra-prediction modes for 16×16 blocks are supported in H.264. Four intra-prediction modes for 8×8 blocks of the chrominance components are also supported. The intra 4×4 modes are suitable for coding parts of the video frame with small details. On the contrary, prediction of the whole 16×16 block is more suited to coding smooth areas of the frame. Another prediction mode for the luminance component is called I-PCM. It is provided mainly for coding anomalous image content. In this mode, pixels are transmitted without any prediction or transformation.

A 4×4 prediction mode specifies a way to generate 16 predictive pixel values $B(i, j), i = 0, 1, 2, 3, j = 0, 1, 2, 3$, using some or all of the neighboring pixels $P(i, j)$, where $i = -1, j = -1, 0, \ldots, 6$, or $i = -1, 0, \ldots, 3, j = -1$. The pixel labeling is shown in Fig. 9.6.

P(−1,−1)	P(−1,0)	P(−1,1)	P(−1,2)	P(−1,3)	P(−1, 4)	P(−1, 5)	P(−1, 6)
P(0,−1)	B(0,0)	B(0,1)	B(0,2)	B(0,3)			
P(1,−1)	B(1,0)	B(1,1)	B(1,2)	B(1,3)			
P(2,−1)	B(2,0)	B(2,1)	B(2,2)	B(2,3)			
P(3,−1)	B(3,0)	B(3,1)	B(3,2)	B(3,3)			

Figure 9.6   Pixel labeling used by 4×4 intra prediction modes

Formulas for calculating the predicted values for the nine 4×4 intra-prediction modes are tabulated in Table 9.1.

Mode 0 is a *vertical* prediction mode in which each pixel of the jth column is predicted by the same value $P(-1, j)$. Mode 1 is a *horizontal* prediction mode in which each pixel of the ith row is predicted by the same value $P(i, -1)$. Mode 2 is a *DC* prediction mode in which all pixels of the current 4×4 block are predicted by the same value equal to the arithmetic mean of eight previously processed values. Modes 3–8 are called *diagonal-down-left*, *diagonal-down-right*, *vertical-right*, *horizontal-down*, *vertical-left*, and *horizontal-up*, respectively. The main idea behind these *diagonal* prediction modes is to improve the prediction of differently directed edges in a video frame.

For regions with fewer spatial details, typically 16×16 intra-prediction modes are used. There are four 16×16 prediction modes: *vertical*, *horizontal*, *DC*, and *plane* modes. The first three modes are specified similarly to the 4×4 prediction modes except for the numbers of neighboring pixels. The plane mode is specified as follows:

$$B_{i,j} = \left[(a + b(i - 7) + c(j - 7)) / 32 \right], \, i, j = 0, 1, \ldots, 15$$

where $B_{i,j}$ is clipped to the range $[0, 255]$,

$$a = 16 \left(P(-1, 15) + P(15, -1) \right), b = \left[5H/64 \right], c = \left[5V/64 \right]$$

$$H = \sum_{i=0}^{7} (i + 1) \left(P(8 + i, -1) - P(6 - i, -1) \right)$$

$$V = \sum_{j=0}^{7} (j + 1) \left(P(-1, 8 + j) - P(-1, 6 - j) \right)$$

and $[\cdot]$ denotes rounding off. In this mode the prediction is a plane with a luminance gradient depending on the luminance gradient in the neighboring row $P(-1, j)$, $j = 0, 1, \ldots, 15$, and in the neighboring column $P(i, -1)$, $i = 0, 1, \ldots, 15$. The mode is suitable for areas with slowly varying luminance. The four supported 8×8 intra-prediction modes for chrominance components are similar to the 16×16 prediction modes for the luminance component.

Compared to the previous standards, the H.264 standard is rather flexible in the selection of block sizes for temporal prediction. *Inter frame prediction* in H.264/AVC is based on *hierarchical splitting of a 16×16 macroblock into blocks of smaller sizes*. In particular, a 16×16 macroblock of the luminance component can be used without splitting or it can be split either into two 16×8 or 8×16 blocks or into four 8×8 blocks. If the 8×8 partition is chosen, then each of the 8×8 blocks can be split in turn, either into two 8×4 or 4×8 subblocks or into four 4×4 subblocks. At most 16 motion vectors can be transmitted for each 16×16 macroblock. Typically, due to this partition, smaller blocks are used for moving objects and larger blocks for background. This

Table 9.1 4×4 intra-prediction modes

Mode	Prediction	Weighting coefficients
0	$B_{i,j} = P_{-1,j}$	
1	$B_{i,j} = P_{i,-1}$	
2	$B_{i,j} = \left[\left(\sum_{i=0}^{3} P_{i,-1} + \sum_{j=0}^{3} P_{-1,j} \right) / 8 \right]$	
3	$B_{i,j} = \left[\left(\sum_{k=0}^{2} w_k P_{-1,i+j+k} \right) / 4 \right]$	$w_0 = 1$ $w_1, w_2 = \begin{cases} 3, 0, & i = j = 3, \\ & P_{-1,7} \text{ is available} \\ 2, 1, & \text{otherwise} \end{cases}$
4	$B_{i,j} = \begin{cases} \left[\dfrac{\sum_{k=0}^{2} w_k P_{i-j-k,-1}}{4} \right], & i > j \\[2mm] \left[\dfrac{\sum_{k=0}^{2} w_k P_{-1,j-i-k}}{4} \right], & i < j \\[2mm] \left[\dfrac{P_{0,-1} + 2P_{-1,-1} + P_{-1,0}}{4} \right], & i = j \end{cases}$	$w_k = (1, 2, 1)$
5	$B_{i,j} = \begin{cases} \left[\dfrac{\sum_{k=0}^{2} w_k P_{i-\lfloor j/2 \rfloor -k,-1}}{4} \right], & z_{\mathrm{VR}} \geq 0 \\[2mm] \left[\dfrac{P_{-1,0} + 2P_{-1,-1} + P_{0,-1}}{4} \right], & z_{\mathrm{VR}} = -1 \\[2mm] \left[\dfrac{P_{-1,j-1} + 2P_{-1,j-2} + P_{-1,j-3}}{4} \right], & z_{\mathrm{VR}} = -2, -3 \end{cases}$ $z_{\mathrm{VR}} = 2i - j$	$w_k = \begin{cases} (2, 2, 0), & z_{\mathrm{VR}} \text{ is even} \\ (1, 2, 1), & z_{\mathrm{VR}} \text{ is odd} \end{cases}$
6	$B_{i,j} = \begin{cases} \left[\dfrac{\sum_{k=0}^{2} w_k P_{-1,j-\lfloor i/2 \rfloor -k}}{4} \right], & z_{\mathrm{HD}} \geq 0 \\[2mm] \left[\dfrac{P_{-1,0} + 2P_{-1,-1} + P_{0,-1}}{4} \right], & z_{\mathrm{HD}} = -1 \\[2mm] \left[\dfrac{P_{i-1,-1} + 2P_{i-2,-1} + P_{i-3,-1}}{4} \right], & z_{\mathrm{HD}} = -2, -3 \end{cases}$ $z_{\mathrm{HD}} = 2j - i$	$w_k = \begin{cases} (2, 2, 0), & z_{\mathrm{HD}} \text{ is even} \\ (1, 2, 1), & z_{\mathrm{HD}} \text{ is odd} \end{cases}$
7	$B_{i,j} = \left[\dfrac{\sum_{k=0}^{2} w_k P_{-1,i+\lfloor j/2 \rfloor +k}}{4} \right]$	$w_k = \begin{cases} (1, 1, 0), & j \text{ is even} \\ (1, 2, 1), & j \text{ is odd} \end{cases}$
8	$B_{i,j} = \begin{cases} \left[\dfrac{\sum_{k=0}^{2} w_k P_{j+\lfloor i/2 \rfloor +k,-1}}{4} \right], & z_{\mathrm{HU}} < 5 \\[2mm] \left[\dfrac{P_{2,-1} + P_{3,-1}}{4} \right], & z_{\mathrm{HU}} = 5 \\[2mm] P(3, -1), & z_{\mathrm{HU}} > 5 \end{cases}$ $z_{\mathrm{HU}} = i + 2j$	$w_k = \begin{cases} (2, 2, 0), & z_{\mathrm{HU}} \text{ is even} \\ (1, 2, 1), & z_{\mathrm{HU}} \text{ is odd} \end{cases}$

increases the quality of the synthesized video and reduces computational complexity of the motion compensation procedure.

The standard allows us to use *multiple reference frames*. It means that P and B-type predictions are performed by using a number of previous and future (in display order) frames, not just the most recent or the next frame. The possibility for the encoder to search for the best prediction among a larger number of already decoded and stored frames is important; for example, when the camera switches between two scenes or there exists other periodicity in video sequence. For P-type predictions a list of already decoded past frames is used. The B-type prediction is generalized and in this standard it denotes *bipredictive* instead of *bidirectional* prediction. For B-type predictions, besides a list of already decoded past frames, one additional list of already decoded future frames is used. In this mode two temporal predictors of a given block are constructed. They can be chosen from two past reference frames, from two future reference frames, or from one reference frame in the past and one reference frame in the future. Notice that in H.264 a B-frame can be predicted from B-frames. A linear combination of these two predictors with arbitrary weights is then used as the prediction for the current block. For each block of size 16×16, 16×8, 8×16, or 8×8, if the corresponding partition is chosen, a different reference frame can be used. If an 8×8 block is further split into smaller subblocks, then the same reference frame is used to search for the best prediction of all subblocks.

Macroblocks can also be coded in the so-called skip mode. In this case no prediction error and motion vectors are transmitted. P and B skip modes are supported. Unlike the previous standards, *motion vectors* (MVs) *are predicted (differentially coded)*. It is taken into account that MVs of the adjacent macroblocks (blocks or subblocks) tend to be highly correlated. The prediction is calculated as the median value of three neighboring blocks on the left, top, and top-right (or top-left, if top-right is not available). More precisely, for macroblocks of size 16×16 and for blocks of size 8×8, the MV prediction is

$$MV_{\mathrm{p}} = \begin{cases} Median(MV_{\mathrm{A}}, MV_{\mathrm{B}}, MV_{\mathrm{C}}), & \text{if } MV_{\mathrm{C}} \text{ is available} \\ Median(MV_{\mathrm{A}}, MV_{\mathrm{B}}, MV_{\mathrm{D}}), & \text{otherwise} \end{cases}$$

where MV_i is the motion vector of the neighboring block i, $i \in \{A, B, C, D\}$, and

$$Median(x, y, z) = x + y + z - \min(x, y, z) - \max(x, y, z).$$

For blocks of size 8×16, $MV_{\mathrm{p}} = MV_{\mathrm{B}}$ or $MV_{\mathrm{p}} = MV_{\mathrm{A}}$ for the upper and the lower 8×16 blocks, respectively. For blocks of size 16×8, $MV_{\mathrm{p}} = MV_{\mathrm{A}}$ or $MV_{\mathrm{p}} = MV_{\mathrm{C}}$ for the left and right 16×8 blocks, respectively. The difference between the original MV and its prediction is then entropy-coded. Notice that in the skip mode, the MV assigned to the skipped macroblock coincides with its prediction. The key difference compared to other standards is that, even in the skip mode, a nonzero motion of the macroblock is accepted.

In order to increase efficiency of motion estimation, the pixel values are first *interpolated to achieve 1/4th-pixel accuracy* for the luminance component and up to *1/8th accuracy* for chrominance components. Interpolation of the luminance component is

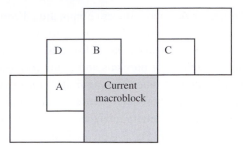

Figure 9.7   MV prediction

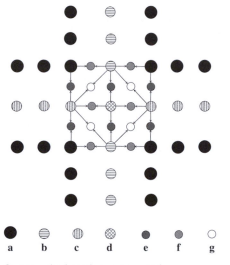

Figure 9.8   Quarter-pixel motion compensation

started with half-pixel interpolation performed by applying a 6-tap FIR filter with the following coefficients: $1/32, -5/32, 20/32, 20/32, -5/32, 1/32$. As shown in Fig. 9.8, type "b" half-pixel values are obtained by filtering six adjacent integer-position (type "a") pixels horizontally. Type "c" half-pixel values are obtained by filtering the corresponding six type "a" pixels vertically. Type "d" half-pixel values can be calculated by filtering the obtained six type "c" half-pixel values horizontally or type "b" half-pixel values vertically. Type "e," "f," and "g" quarter-pixel values are obtained by averaging two adjacent type "a," "b," "c," or "d" pixels of integer-pixel or half-pixel accuracy. A 1/8th-pixel accuracy for chrominance components is provided similarly by linear interpolation of the adjacent integer-position pixels.

Additional improvement in inter-prediction is related to the so-called *weighted prediction* and using a *deblocking filter in the prediction loop*. Weighting implies that for B-type prediction, instead of averaging two motion-compensated predictions, their linear combination with arbitrary coefficients can be used. Moreover, motion-compensated predictions can be offset by values specified by the encoder. Weighted prediction is allowed even for P-type prediction. This improves prediction for areas where one scene fades into another. The deblocking filter incorporated into the prediction loop reduces blocking artifacts in reference frames used for prediction. This leads to the better sub-jective quality of the reconstructed frames and reduces the bit-rate, since the block boundaries become smoother and can be represented by fewer bits.

One of the main improvements is related to the *transform and quantization* used in H.264. Similar to other video-coding standards, H.264 exploits a transform coding technique. Notice that for I-frames, the transform is applied to the intra-prediction resid-ual and for P and B-frames to the inter-prediction residual. However, the transform is applied to 4 × 4 blocks and instead of a 4 × 4 DCT, a separable orthogonal integer trans-form is used. This transform is derived from the 4 × 4 DCT-II transform as follows. Let T be the transform matrix of the 4 × 4 DCT-II; that is,

$$T = \begin{pmatrix} a & a & a & a \\ b & c & -c & -b \\ a & -a & -a & a \\ c & -b & b & -c \end{pmatrix} \tag{9.2}$$

where $a = 1/2$, $b = \sqrt{1/2}\cos(\pi/8)$, and $c = \sqrt{1/2}\cos(3\pi/8)$.

Then, the 2D-transform of the matrix X of size 4 × 4 can be represented in the form

$$Y = TXT^{\mathrm{T}} = DAXA^{\mathrm{T}}D \tag{9.3}$$

where

$$A = \begin{pmatrix} 1 & 1 & 1 & 1 \\ 1 & d & -d & -1 \\ 1 & -1 & -1 & 1 \\ d & -1 & 1 & -d \end{pmatrix}$$

$$D = \begin{pmatrix} a & 0 & 0 & 0 \\ 0 & b & 0 & 0 \\ 0 & 0 & a & 0 \\ 0 & 0 & 0 & b \end{pmatrix}$$

and where $d = c/b \approx 0.414$. In order to simplify the implementation of the transform, d is approximated by 0.5. To keep the transform orthogonal, b is approximated by $\sqrt{2/5}$. For further simplification, the transform can be rewritten in the following equivalent form

$$Y = (AXA^{\mathrm{T}}) \otimes B \tag{9.4}$$

where

$$B = \begin{pmatrix} a^2 & ab & a^2 & ab \\ ab & b^2 & ab & b^2 \\ a^2 & ab & a^2 & ab \\ ab & b^2 & ab & b^2 \end{pmatrix}$$

and $\otimes$ denotes a componentwise multiplication. After scaling the second and fourth rows of matrix A by 2 and the same rows of matrix D by 1/2, we obtain the final form of the transform

$$Y = A_N X A_N^{\mathrm{T}} \otimes B_N \tag{9.5}$$

where

$$A_N = \begin{pmatrix} 1 & 1 & 1 & 1 \\ 2 & 1 & -1 & -2 \\ 1 & -1 & -1 & 1 \\ 1 & -2 & 2 & -1 \end{pmatrix}$$

and

$$B_N = \begin{pmatrix} a^2 & \frac{ab}{2} & a^2 & \frac{ab}{2} \\ \frac{ab}{2} & \frac{b^2}{4} & \frac{ab}{2} & \frac{b^2}{4} \\ a^2 & \frac{ab}{2} & a^2 & \frac{ab}{2} \\ \frac{ab}{2} & \frac{b^2}{4} & \frac{ab}{2} & \frac{b^2}{4} \end{pmatrix}.$$

The inverse of (9.5) has the form

$$X = CDYDC^{\mathrm{T}} \tag{9.6}$$

where

$$C = \begin{pmatrix} 1 & 1 & 1 & 1/2 \\ 1 & 1/2 & -1 & -1 \\ 1 & -1/2 & -1 & 1 \\ 1 & -1 & 1 & -1/2 \end{pmatrix}$$

or

$$X = C\,(Y \otimes B)\,C^{\mathrm{T}}. \tag{9.7}$$

It is easy to check that the transform (9.5) is an orthogonal transform, since $(D_N A)^{-1} = (D_N A)^{\mathrm{T}} = (CD)$, where

$$D_N = \begin{pmatrix} a & 0 & 0 & 0 \\ 0 & b/2 & 0 & 0 \\ 0 & 0 & a & 0 \\ 0 & 0 & 0 & b/2 \end{pmatrix}.$$

The transform (9.5) is not identical to the DCT-II 4×4 transform. However, it has almost the same efficiency as the DCT-II transform from the compression point of view and it has a number of important properties simplifying its implementation:

- The multiplication by the matrix A_N in (9.5) can be implemented in integer arithmetic by using only additions, subtractions, and shifts.
- The inverse of (9.5) is an integer transform. Unlike the DCT-II and IDCT-II transforms, the modified transform (9.5) and its inverse can be implemented in integer arithmetic without any loss of accuracy.

Multiplication by matrices B_N and B in (9.5) and (9.7), respectively, is combined with the scalar quantization (dequantization). In other words, in H.264 post-scaling by B_N and quantization are performed as follows

$$Z_{i,j} = \text{sgn}(Y_{i,j}) \left\lfloor \frac{|Y_{i,j}|}{Qstep} + \Delta(Qstep) \right\rfloor \tag{9.8}$$

$$= \text{sgn}(P_{i,j}) \left\lfloor \frac{|P_{i,j}|S_{i,j}}{Qstep} + \Delta(Qstep) \right\rfloor, i = 1,2,\ldots,4, \ j = 1,2,\ldots,4 \tag{9.9}$$

where $\lfloor x \rfloor$ denotes the closest integer less than or equal to x, $\Delta(Qstep)$ is a parameter which controls the extension of the quantization cell with zero approximating value (zero zone), $Qstep$ is a quantization step, $Y_{i,j}$ is a transform coefficient, $P_{i,j}$ is the corresponding coefficient of the core transform $P = A_N X A_N^\mathrm{T}$, and $S_{i,j} \in \{a^2, b^2/4, ab/2\}$ is the corresponding scaling factor from B_N. The quantization step $Qstep$ is a function of the quantization parameter QP which takes on 52 values. The quantization step increases two times for every increment of 6 in QP and can be represented as follows

$$Qstep(QP) = q_k 2^{\lfloor QP/6 \rfloor}, k = QP \bmod 6$$

where $q_0 = 0.625$, $q_1 = 0.6875$, $q_2 = 0.8125$, $q_3 = 0.875$, $q_4 = 1$, and $q_5 = 1.125$. In order to implement (9.9) in integer arithmetic without division, it is modified as follows,

$$Z_{i,j} = \text{sgn}(Y_{i,j}) \left(|P_{i,j}| M_{i,j}(QP) + \lfloor \alpha \rfloor \right) >> \beta \tag{9.10}$$

where

$$M_{i,j}(QP) = \frac{2^{15} S_{i,j} 2^{\lfloor QP/6 \rfloor}}{Qstep(QP)} = \frac{2^{15} S_{i,j}}{q_{QP\bmod 6}}$$

$$\beta = 15 + \lfloor QP/6 \rfloor$$

$$\alpha = \begin{cases} 2^\beta/3 & \text{for intra blocks} \\ 2^\beta/6 & \text{for inter blocks} \end{cases}$$

and $>>$ denotes a binary shift right. It is easy to see that for each value of the scaling coefficient S there exist only six different values of the modified scaling coefficient M that are tabulated in the standard.

The dequantization is combined with prescaling by coefficients from the matrix B and scaling by a constant factor 64 which is used in order to avoid rounding errors

$$W_{i,j} = \hat{Y}_{i,j} S_{i,j} 64 \tag{9.11}$$

where

$$\hat{Y}_{i,j} = Z_{i,j} Qstep(QP).$$

Formula (9.11) can be rewritten as

$$W_{i,j} = Z_{i,j} V_{i,j}(QP) 2^{\lfloor QP/6 \rfloor}$$

where

$$V_{i,j}(QP) = S_{i,j}64q_{\text{QPmod}6}$$

is tabulated for each value of scaling coefficient S. Then, the reconstructed and scaled coefficient $W_{i,j}$ is transformed by the core inverse transform and divided by 64; that is, implemented as a binary right shift

$$\hat{X} = CWC^{\text{T}} >> 6$$

where $W = \{W_{i,j}\}, i = 1, 2, \ldots, 4, j = 1, 2, \ldots, 4.$

Since in the H.264 standard some macroblocks can also be coded in 16×16 intra mode for such blocks, 2-stage transform is applied. The first stage is the core transform of (9.5) performed for each of four 4×4 subblocks of the macroblock. The obtained four DC coefficients are grouped into 4×4 block Y_{DC} and further transformed by using a 4×4 Hadamard transform

$$U_{\text{DC}} = HY_{\text{DC}}H^{\text{T}}/2$$

where

$$H = \begin{pmatrix} 1 & 1 & 1 & 1 \\ 1 & 1 & -1 & -1 \\ 1 & -1 & -1 & 1 \\ 1 & -1 & 1 & -1 \end{pmatrix}.$$

The obtained matrix U_{DC} is quantized according to (9.10) with

$$M_{i,j} = M_{0,0} = \frac{2^{15}a^2}{q_{\text{QPmod}6}} = \frac{2^{13}}{q_{\text{QPmod}6}}$$

and

$$\beta = 16 + 2^{\lfloor QP/6 \rfloor}.$$

At the decoder, first the inverse Hadamard transform is applied to the matrix of the obtained quantized DC coefficients Z_{DC},

$$W_{\text{DC}} = HZ_{\text{DC}}H^{\text{T}}.$$

The output matrix W_{DC} is then scaled as follows,

$$WS_{i,j} = W_{\text{DC}(i,j)}V_{0,0}2^{\lfloor QP/6 \rfloor - 2}, QP \geq 12$$

or

$$WS_{i,j} = \left(W_{\text{DC}(i,j)}V_{0,0} + 2^{1 - \lfloor QP/6 \rfloor}\right) >> (2 - \lfloor QP/6 \rfloor), QP < 12$$

where

$$V_{0,0}(QP) = a^2 64q_{\text{QPmod}6} = 16q_{\text{QPmod}6}.$$

The scaled coefficients WS are placed into the corresponding 4×4 subblocks of the 16×16 block. Then, the inverse core transform is performed for each of the subblocks. The DC coefficients of 4×4 blocks of the chrominance components are grouped into blocks of size 2×2 and processed by using a 2×2 Hadamard transform described by the following matrix

$$H = \begin{pmatrix} 1 & 1 \\ 1 & -1 \end{pmatrix}.$$

The transform coefficients are quantized in the same way as the luminance component. At the decoder, first the inverse Hadamard transform is performed and then the obtained coefficients are scaled as follows,

$$WS_{i,j} = W_{DC(i,j)} V_{0,0} 2^{\lfloor QP/6 \rfloor - 1}, QP \geq 6$$

or

$$WS_{i,j} = \left(W_{DC(i,j)} V_{0,0} \right) >> 1, QP < 6.$$

The scaled DC coefficients are inserted into the corresponding 4×4 blocks and are transformed by the inverse core 4×4 transform.

Entropy coding of the transform coefficients has two modes: Context-based Adaptive Variable Length Coding (CAVLC) and Context-based Adaptive Binary Arithmetic Coding (CABAC).

CAVLC is the variable-length coding method with the reduced computational complexity. It is applied to the zigzag scanned 4×4 block of the transformed coefficients. Encoding of the 16 transform coefficients is reduced to encoding the following parameters of the zigzag ordered block:

- the number of nonzero coefficients (Nz) and number of ± 1s called *Trailing ones* ($Tr1s$) if it is less than or equal to 3. These two values are coded jointly. Notice that $Tr1s$ is counted starting with the last highest-frequency nonzero coefficient towards the lowest-frequency nonzero coefficient; that is, in reverse order;
- the sign of each of $Tr1s$ trailing ones (S_i, $i = 1, 2, \ldots, Tr1s$);
- the amplitudes and signs ($Level_i$, $i = 1, 2, \ldots, Nz - Tr1s$) of the other than trailing ones, $Nz - Tr1s$ nonzero coefficients;
- the total number of zeros occurring before the last nonzero coefficient (Tz);
- the number of zeros occurring before each nonzero coefficient in the zigzag scanned block of the transform coefficients represented in the reverse order (Run_i, $i = 1, 2, \ldots, Nz$).

The important feature of CAVLC is the context-adaptivity. For each of the parameters enumerated above, a few predetermined lookup VLC tables are used. The choice of the tables depends on the already encoded parameters of either neighboring or current 4×4 coefficient blocks. Since the numbers of nonzero coefficients in the neighboring blocks are correlated, the choice of one of four VLC tables for coding the pair ($Nz, Tr1s$) depends on the number of nonzero coefficients in previously coded upper and left-hand blocks (Nz_u and Nz_l, respectively). More precisely, the parameter Nz_c, which controls the choice of the table, is calculated as follows:

$$Nz_c = \begin{cases} \left[\frac{Nz_u + Nz_l}{2}\right] & \text{if both blocks are available} \\ Nz_u & \text{if only the upper block is available} \\ Nz_l & \text{if only the left-hand block is available} \\ 0 & \text{otherwise.} \end{cases}$$

Notice that not more than three trailing ones, counting from the last nonzero coefficient, are encoded as a special case, that is, jointly with Nz. The signs of $Tr1s$ trailing ones are encoded by one bit each (0 is used for $+$ and 1 is used for $-$). The amplitudes and signs of $Nz - Tr1s$ nonzero coefficients are coded jointly by using exponential Golomb codes (see Appendix). In order to take into account the signs of the coefficients, first the parameter *codenum* for each coefficient value l is computed as follows

$$codenum = \begin{cases} 2|l| - 2 & l > 0 \\ 2|l| - 1 & l < 0. \end{cases}$$

Then, the value m of the Golomb code is selected depending on the previously coded nonzero coefficient. If this coefficient is the first, then m is set to 0, otherwise the previously coded coefficient is compared with a threshold. If it exceeds the threshold, then m is increased by 1. The codeword of the Golomb code with given m for a given value *codenum* consists of two parts: prefix and suffix, where

$$prefix = unar\left(\left\lfloor\frac{codenum}{T}\right\rfloor + 1\right)$$

$$suffix = codenum \mod T$$

$T = 2^m$, $unar(x)$ denotes the unary code of x, that is, the run of $x - 1$ zeros followed by 1; the $suffix$ has length m.

For example, let $l = -2$ and $m = 1$, then $codenum = 2|l| - 1 = 2 \times 2 - 1 = 3$, $prefix = unar\left(\lfloor\frac{3}{2^1}\rfloor + 1\right) = unar(2) = 01$ and $suffix = (3 \mod 2) = 1$. Thus, for $l = -2$ we obtain the codeword 011. However, if $m = 3$ for the same level value $l = -2$, we obtain $prefix = unar(0 + 1) = 1$ and $suffix = (3 \mod 8) = 3$, that is, the corresponding codeword is 1011. In total, six Golomb code tables are available in the standard. The last two entries in each table can be considered as escape codes. The escape code followed by a uniform code of the level is used to encode the levels above the last regularly coded level in a table.

If the number of nonzero coefficients Nz is already known, then it determines the maximum value of the sum of all zeros preceding the last nonzero coefficient Tz. For example, if $Nz = 4$, then $0 \le Tz \le 12$. In total, 15 tables are available for $1 \le Nz \le 15$. If $Nz = 16$, then Tz is not coded since it is known to be zero. The number of zeros occurring before each nonzero coefficient in the zigzag scan Run_i, $i = 1, 2, \ldots, Nz$, is encoded starting with the last nonzero coefficient. The parameters Run_i, $i = 1, 2, \ldots, Nz$, show how the Tz zeros are distributed between the

-2	5	-1	0
2	0	-1	0
1	0	0	0
-1	0	0	0

Figure 9.9 4×4 transformed block

nonzero coefficients. In order to save bits for the representation of the zero runs, the variable-length code for each run is selected depending on the number of zeros which have not yet been coded ($Zerosleft$). For example, if $Zerosleft = 3$, it means that the next run has to be in the range $0-3$. There are seven code tables in the standard for $Zerosleft = 1, 2, 3, 4, 5, 6, \geq 6$. We start encoding with $Zerosleft = Tz$ and then, when the ith zero run Run_i is encoded, the number of zeros is updated as $Zerosleft \leftarrow Zerosleft - Run_i$. If $Zerosleft = 0$, it means that the next zero runs have zero lengths and it is not necessary to encode them. Neither is it necessary to encode Run_{Nz}, that is, the run before the lowest-frequency coefficient.

Example 9.1   Consider the 4×4 transformed block shown in Fig. 9.9. Assume that $Nz_c = 0$.

The zigzag scanned block has the form: $-2, 5, 2, 1, 0, -1, 0, -1, 0, -1, 0, 0, 0, 0, 0, 0$.

For encoding the pair $(Nz, Tr1s) = (7, 3)$ we choose the VLC0 table. The corresponding codeword is 00010100.

The signs of trailing ones $S_1 = \text{``}-\text{''}, S_2 = \text{``}-\text{''}, S_3 = \text{``}-\text{''}$ are encoded as 111.

For encoding the $Level_1 = +1$, we use the Golomb code with $m = 0$. We compute $codenum = 2 \times 1 - 2 = 0$, $prefix = unar(\lfloor 0/1 \rfloor + 1) = 1$, the suffix in this case has zero length. The codeword is 1. The next level $Level_2 = 2$ is encoded by the Golomb code with $m = 1$. We increase m, since the previous coefficient is greater than 0. The corresponding parameter $codenum = 2$ is encoded by the codeword with $prefix = unar(\lfloor 2/2 \rfloor + 1) = 01$ and $suffix = 2 \mod 2 = 0$. The codeword is 010. The coefficient 5 we encode by the same Golomb code. The corresponding codeword is 000010. Since this coefficient is greater than 3, we increase m by one and encode -2 by the Golomb code with $m = 2$. We obtain $codenum = 3$, $prefix = unar(\lfloor 3/4 \rfloor + 1) = 1$, and $suffix = 3 \mod 4$. The corresponding codeword is 111.

The total number of zeros between the nonzero coefficients is $Tz = 3$. We initiate $Zerosleft$ by 3 and encode $Run_7 = 1$ by 00. Then $Zerosleft$ is decreased by 1 and $Run_6 = 1$ is encoded by 01. We again decrease $Zerosleft$ by 1 and encode $Run_5 = 1$ by 0. After this step $Zerosleft$ becomes equal to 0. Since there are no zeros left, it is not necessary to encode $Run_1 - Run_4$.

The second coding mode called CABAC is typically more efficient compared to CAVLC but has larger computational complexity. This mode is based on binary arithmetic coding. The arithmetic coding is a kind of block lossless coding. As shown in the Appendix, it has significantly less redundancy than any kind of symbol-by-symbol lossless coding. The difference in redundancy with respect to symbol-by-symbol lossless coding is especially noticeable if the entropy of the data source is much less than 1 bit/sample as it often takes place for prediction residuals. Rather high computational complexity of nonbinary arithmetic coding is usually considered as the main shortcoming of this type of lossless coding. This explains the usage of binary arithmetic coding having significantly lower computational complexity. Thus, a nonbinary input stream is first converted into a binary stream. This conversion can be done in different ways depending on the used binarization scheme. In the H.264 standard the following four binarization schemes and their combinations are used:

- unary binarization: an unsigned integer is coded by the unary code, i.e. $x \geq 0$ is encoded as $unar(x+1)$;
- truncated unary binarization: it is defined only for $0 \leq x \leq M$; if $x = M$, then the terminating "1" is omitted;
- kth-order Golomb binarization: an unsigned integer is coded by the Golomb code with a given parameter k, i.e. $T = 2^k$;
- fixed-length binarization: an unsigned integer $0 \leq x < M$ is given by its binary representation of length $\lceil \log_2 M \rceil$ bits.

An important requirement to any binarization scheme is that the entropy of the obtained binary source has to coincide with the entropy of the original nonbinary source. To satisfy this requirement, proper conditional probabilities have to be assigned to each symbol of the binary data stream. A detailed description of the binarization process can be found in the Appendix.

The next step after binarization is context modeling. However, this step is used only in the so-called *regular coding* mode. Another, the so-called *bypass coding* mode, is applied without this step. Context modeling implies that each symbol of the obtained binary data stream is classified depending on a value of a chosen context; that is, a predefined set of already processed neighboring symbols. A chosen context model determines a probability distribution which will be used by the arithmetic coder in order to encode the corresponding symbol. The main goal of context modeling is to achieve better compression through context conditioning. Usually, the possibility of context modeling on a subsymbol (bit) level is considered as an advantage of CABAC. This type of context modeling uses relatively few different context values but at the same time takes into account rather high-order conditional probabilities. The bypass coding mode is used for those symbols which are assumed to be nearly uniformly distributed. For such symbols, this simplified mode enables the encoding procedure to be speeded up.

The last step of the CABAC procedure is binary arithmetic coding. Since the multiplication operation is the most difficult and time-consuming operation, CABAC is based on a multiplication-free version of the arithmetic coder similar to the QM-coder described in the Appendix. The adaptation of probability distributions corresponding to different context values is implemented as a prediction algorithm. This algorithm exploits a finite-state machine to update symbol probabilities. The finite-state machine for each context model in CABAC can be in one of 128 states. Each state corresponds to one of 64 possible probabilities of the Less Probable Symbol (LPS) and the value of the More Probable Symbol (MPS) being either 0 or 1. Transitions between the states are determined by the input context value and the value of the input symbol. If a value predicted from the context coincides with the input symbol value, then such a symbol is classified as MPS. Otherwise, the input symbol is classified as LPS. Although 64 probabilities corresponding to states of the finite-state machine are tabulated, they approximate solutions of the recurrent equation

$$p_\sigma = \alpha p_{\sigma-1}$$

where

$$\alpha = \left(\frac{0.01875}{0.5}\right)^{1/63}$$

$p_0 = 0.5$ and $\sigma = 0, 1, \ldots, 63$. Such an empiric choice of parameters provides rather fast adaptation of the symbol probability and, on the other hand, gives sufficiently accurate probability estimates.

Updating the LPS probability is performed according to the formula

$$P(x_{t+1} = \text{LPS}) = \begin{cases} \max\{\alpha P(x_t = \text{LPS}), p_{62}\} & \text{if MPS occurs} \\ \alpha P(x_t = \text{LPS}) + (1 - \alpha) & \text{if LSP occurs} \end{cases} \tag{9.12}$$

where equality implies that the left-hand side and the right-hand side values are rounded to the same tabulated probability value. In fact, instead of computations according to (9.12), the arithmetic encoder simply uses the corresponding state indices σ.

Notice that really only 126 states are used in the probability adaptation process. The probability p_{63} corresponds to a nonadaptive state with a fixed value of MPS and is reserved for encoding of binary decisions before terminating the arithmetic codeword. Following from (9.12), if a MSP enters the encoder, then the previous state index is increased by one (LSP probability reduces) until it is less than 62. If the state with index $\sigma = 62$ is reached, the encoder remains in this state until a LPS occurs. The state with index $\sigma = 62$ corresponds to the minimal LSP probability and the maximal MPS probability. When a LPS enters the encoder in the state with index $\sigma = 62$ the probability "jumps up" over a few states. In general, each LPS occurrence corresponds to a decreasing state index (increasing the LPS probability) by a number, depending on the previous state of the finite-state machine. If the state with $\sigma = 0$ which corresponds to $p_0 = 0.5$ is reached, then the state index is not changed but the values of the LPS and MPS are interchanged.

As well as MPEG-4, the H.264 standard includes sets of profiles and levels which provide for a high level of flexibility of this standard.

Problems

9.1 Convert a given sequence of 30 video frames represented in the BMP format into sequences of the Y, U, and V components.

9.2 For the first frame in the sequence of Y components obtained when solving Problem 9.1, perform DCT-II by blocks of size 8×8, uniformly scalar quantize the obtained coefficients, and reconstruct the frame from the quantized coefficients by applying the IDCT-II. Recommendation: quantize the DC coefficient of each 8×8 block by quantization step 8 and use another quantization step for all AC coefficients of the block.

9.3   Perform motion compensation for the sequence of Y components obtained when solving Problem 9.1. Apply the block matching motion compensation technique for macroblocks of size 16×16. Use for prediction only the previously reconstructed frames. Compute the motion vectors for the macroblocks. Compute the sequence of difference frames.

9.4   For the sequence of difference frames obtained when solving Problem 9.3, perform DCT-II by blocks of size 8×8, uniformly scalar quantize the obtained transform coefficients.

9.5   Perform run-length coding of the transformed first frame from Problem 9.2 and the sequence of difference frames from Problem 9.3. Estimate the entropy of the sequence of runs. Estimate the entropy of the sequence of levels. Estimate the entropy of the motion vectors. Estimate the total number of bits for the quantized coefficients and motion vectors.

9.6   For a given video frame represented in the BMP format, perform the YUV transform. For the obtained component Y, perform intra-prediction by blocks of size 4×4. Check which of the nine prediction modes used in the H.264 standard is the best for each of the blocks.

9.7   For a given video frame represented in the BMP format, perform the YUV transform. For the obtained component Y perform the integer DCT-II-like transform from the standard H.264 for blocks of size 4×4. Reconstruct the Y component.

9.8   Assume that the zigzag scanned 4×4 block of the quantized transform coefficients has the form: $3,2,-2,4,1,-1,0,0,0,-1,1,0,0,1,0,0$. Encode the sequence of coefficients by using CAVLC from H.264.

10 Audio-coding standards

The main application of audio compression systems is to obtain compact digital representations of high-quality (CD-quality) wideband audio signals. Typically, audio signals recorded on CDs and digital audio tapes are sampled at 44.1 or 48 kHz and each sample is represented by a 16-bit integer; that is, the uncompressed two-channel stereo CD-quality audio requires $2 \times 44(48) \times 16 = 1.41(1.54)$ Mb/s for transmission. Unlike speech compression systems, the audio codecs process sounds generated by arbitrary sources and they cannot exploit specific features of the input signals. However, almost all modern audio codecs are based on a model of the human auditory system. The key idea behind the so-called *perceptual coding* is to remove the parts of the input signal which the human cannot perceive. The imperceptible information removed by the perceptual coder is called the *irrelevancy*. Since, similarly to speech signals, audio signals can be interpreted as outputs of sources with memory, then perceptual coders remove both irrelevancy and redundancy in order to provide the lowest bit-rate possible for a given quality.

An important part of perceptual audio coders is the *psychoacoustic model* of the human hearing. This model is used in order to estimate the amount of quantization noise that is inaudible. In the next section, we consider physical phenomena which are exploited by the psychoacoustic model.

10.1 Basics of perceptual coding

The frequency range with which audio coders deal is approximately from 20 Hz to 20 kHz. The human auditory system has remarkable detection capability with a range (called a dynamic range) of 120 dB from the quietest to the loudest sounds. However, digitized CD-quality audio signals have a dynamic range of about 96 dB because they use a 16-bit sample representation. The absolute threshold of hearing $T_q(f)$ characterizes the amount of energy in a pure tone signal such that it can be detected by a listener in a noiseless environment. This absolute threshold is computed as

$$
\begin{aligned}
T_q(f) = {} & 3.64(f/1000)^{-0.8} - 6.5e^{-0.6(f/1000-3.3)^2} \\
& + 10^{-3}(f/1000)^4
\end{aligned}
\tag{10.1}
$$

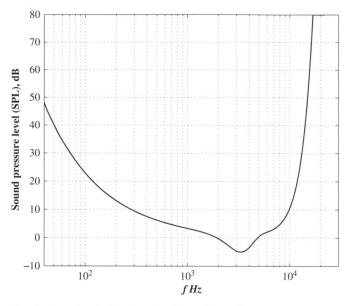

Figure 10.1   The absolute threshold of hearing in a quiet environment

where f denotes a frequency sample (Painter and Spanias 2000). The threshold is measured as a Sound Pressure Level (SPL), normalized by a minimum value, and expressed in dB. The frequency dependence of this threshold is shown in Fig. 10.1.

Using the absolute threshold of hearing in order to estimate the amount of inaudible quantization noise is only the first step in constructing the psychoacoustic model. It transpired that the threshold of hearing at a given frequency depends on the input signal. The reason for that dependency is that louder input sounds can mask or hide weaker ones. The psychoacoustic model takes into account this masking property that occurs when a strong audio signal in the time or frequency domain makes neighboring weaker audio signals imperceptible. The coder analyzes the audio signal and computes the amount of available noise masking as a function of frequency. In other words, the noise detection threshold is a modified version of the absolute threshold $T_q(f)$, with its shape determined by the input signal at any given time.

In order to estimate this threshold, we have first to consider the concept of *critical bands*. Our inner ear (cochlea) can be interpreted as a bank of highly overlapping non-linear (level-dependent) bandpass filters. Different areas of the cochlea, each containing a set of neural receptors, correspond to different frequency bands called *critical bands*. The bandwidths of the cochlea filters (*critical bandwidths*) increase with increasing frequency. They are less than 100 Hz for the lowest audible frequencies and more than 4 kHz at the highest frequencies. It is said that the human auditory system has a limited, frequency-dependent resolution. The critical bandwidth is a function of frequency that quantifies the cochlea filter passbands.

More than 60 years ago, due to experiments by H. Fletcher, it was discovered that a critical band is a range of frequencies over which the masking Signal-to-Noise Ratio (SNR) remains more or less constant. For example, H. Fletcher found that a tone at

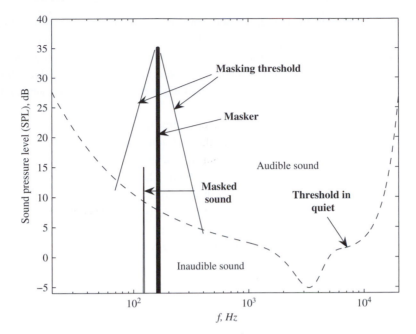

Figure 10.2   Masking of weaker signals by strong tone masker

700 Hz can mask narrowband noise of less energy in the vicinity of approximately ± 70 Hz. If the noise is outside the range 630–770 Hz, then to be inaudible it has to have much less energy than the tone. Thus, the critical bandwidth is 140 Hz. The phenomenon of masking has a simple explanation. A strong tone or noise masker present in a given critical band blocks the detection of a weaker signal, since it creates a strong excitation of the corresponding area of the cochlea. Figure 10.2 illustrates the masking property.

A second observation about masking is that noise and tone have different masking properties. Let B denote a critical-band number and E_T and E_N be the tone and noise masker energy levels, respectively, for this critical band; the following empirical rules were observed:

$$TH_N = E_T - (14.5 + B)\ \text{dB} \tag{10.2}$$

and

$$TH_T = E_N - K\ \text{dB} \tag{10.3}$$

where TH_N and TH_T are the noise-masking thresholds for *tone-masking-noise* and the tone-masking threshold for *noise-masking-tone* types of masking, respectively. The parameter K takes values in the range 3–6 dB. In other words, for the tone-masking-noise case, if the energy of noise (in dB) is less than TH_N, then the noise is imperceptible. For the noise-masking-tone case, if the energy of tone (in dB) is less than TH_T, then the tone is imperceptible. The problem is that speech and audio signals are neither pure tones nor pure noise but rather, a mixture of both. The degree to which a signal, within a critical band, appears more or less tone-like (or noise-like) determines its masking properties. Usually in perceptual coding, masking signals are classified as tone or noise and then the thresholds (10.2) and (10.3) are computed. The masking property of a given masker in the critical band often spreads to the neighboring critical bands and affects detection thresholds in these critical bands. This effect, also known as the spread of masking, or inter-band masking is modeled in coding applications by a *spreading function.*

Since typically each subband of an audio signal to be encoded contains a collection of maskers of tone and noise types, then the individual masking thresholds computed for each of the maskers and modified by applying a spread function are then combined for each subband into a *global masking threshold*. The noise spectrum shaped according to the computed global masking threshold is often called Just Noticeable Distortion (JND). The absolute threshold of hearing $T_q(f)$ is also considered when shaping the noise spectra. Let T_{qmin} denote the minimal value of the absolute threshold of hearing in a given subband; then in this subband max $\{JND, T_{qmin}\}$ is used as the permissible distortion threshold.

Exactly how the human auditory system works is still a topic of active research. For the purpose of constructing perceptual audio coders, all the considered phenomena are described by formulas and exploited to reduce the bit-rate of the perceptual audio coder.

Figure 10.3 is a block diagram of a typical psychoacoustic model of hearing. The input signal in this diagram is actually the short-term signal spectrum. The cochlea filter model computes the short-term cochlea energy model, i.e. a distribution of the energy along the cochlea. This information is used in two ways. The short-term signal spectrum and the cochlea energy for a given subband are used to compute the tonality of the signal in this subband. Tonality is needed because tones mask noise better than noise masks tones. The cochlea energy plus the tonality information allow us to compute the threshold of audibility for the signal in the subband. This is a spectral estimate indicating what noise level will be audible as a function of frequency. As long as the quantization noise is less than the threshold of audibility at all frequencies, the quantization noise will be imperceptible. Figure 10.4 is a diagram for an entire perceptual audio coder. The psychoacoustic model represents one of its blocks. The input signal of the coder is a discrete-time audio signal with sampling rate 32, 44.1, or 48 kHz. Each sample is represented by a 16-bit integer value. The input audio stream is split into frames and for

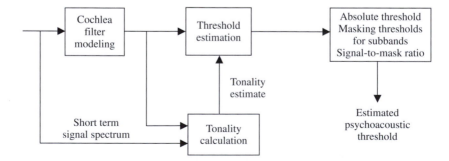

Figure 10.3   Typical psychoacoustic model for audio signal

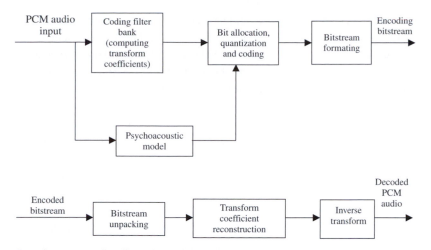

Figure 10.4   Generic perceptual audio coder and decoder

each frame transform coefficients are computed. The audio coder is based on the transform coding technique. From the theoretical point of view any kind of transform such as the DFT, the DCT, or the so-called Modified DCT (MDCT), or the transform based on a filter bank can be used in the first block of the coder. In audio standards the audio stream passes through a filter bank that divides the input into multiple frequency subbands. This special type of transform coding is called *subband coding* (Jayant and Noll 1984). Moreover, as will be shown below, some particular cases of subband coding can be implemented as the MDCT. The distinguishing feature of all transforms used in audio coding is that they are *overlapped transforms*. Using nonoverlapped transforms as, for example, in video coding would lead to audible clicks with block frequency.

The input audio stream simultaneously passes through a psychoacoustic model that determines the *ratio of the signal energy to the masking threshold* (SMR) for each subband. The quantization and coding block uses the SMRs to decide how to allocate the total available number of bits over the subbands to minimize the audibility of the quantization noise. The transform coefficients are then quantized according to the chosen bit allocation based on the perceptual thresholds. Usually, uniform scalar quantization is used. The outputs of the quantizer are then further compressed using lossless coding, most often a Huffman coder. In recent standards, context arithmetic coding is used instead of Huffman coding. The obtained bitstream contains encoded transform coefficients and side information (bit allocation). The decoder decodes this bitstream, restores the quantized subband values, and reconstructs the audio signal from the subband values.

10.2   Overview of audio standards

The most important standards in this domain were developed by the ISO Motion Picture Experts Group (ISO-MPEG) subgroup on high-quality audio. Although perfectly suitable for audio-only applications, MPEG Audio is actually one part of a three-part compression standard that also includes video systems. The original MPEG Audio standard was created for mono sound systems and had three layers, each providing a greater compression ratio. MPEG-2 was created to provide stereo and multichannel audio capability. The MPEG-2 Advanced Audio Coder (MPEG-AAC) duplicates the performance of MPEG-2 at half the bit-rate. Recently, the AC technique was further enhanced and included in the standard MPEG-4 Audio as two efficient coding modes: High Efficiency AAC (HE-AAC) and HE-AAC v2.

The original MPEG-coder is sometimes now referred to as MPEG-1. It contains three independent layers of compression. This provides a wide range of tradeoffs between code complexity and compressed audio quality. Layer I has the lowest complexity and suits best bit-rates above 128 kb/s per channel. For example, Philips' Digital Compact Cassette (DCC), developed in 1992, used layer I compression at 192 kb/s per channel. Layer II has an intermediate complexity and bit-rates around 128 kb/s per channel. The main application for this layer is audio coding for Digital Audio Broadcasting (DAB) and for Digital Video Broadcasting. It uses a more complicated quantization procedure than layer I.

Layer III is the most complex but it provides good audio quality for bit-rates around 64 kb/s per channel. It can be used for transmission audio over ISDN channels and for storing audio on CD. The widely used MP3 player contains a decoder of audio signals compressed according to layer III of the MPEG Audio standard. The specific features of this layer are: hybrid transform coding based on the MDCT and filter bank coding, a more complicated psychoacoustic model, and the so-called echo-suppression system. The variable-length coding of the transform coefficients is used at this layer.

The digital transform implemented as a *polyphase filter bank* is common for layers I and II of the MPEG Audio compression algorithm. This filter bank divides the audio signal into 32 frequency subbands with equal bandwidths. Each subband is 750 Hz wide. Surely, the equal widths of the subbands do not accurately reflect human auditory system frequency-dependent behavior. Since at lower frequencies the critical bandwidth is about 100 Hz, the low-frequency subbands cover more than one critical band. As a result, the critical band with the least noise masking determines the number of bits (number of quantization levels) for the entire subband. At high frequencies, where critical bandwidths are about 4 kHz, one critical band covers a few subbands but this circumstance is less critical in the sense of bit allocation.

Notice that the polyphase filter bank and its inverse are not always lossless transforms. Even without quantization, the inverse transform sometimes cannot perfectly recover the original signal. However, the error introduced by using the filter bank is negligible and inaudible. Adjacent filter bands in this filter bank have a major frequency overlap. A signal at a single frequency can affect two adjacent filter bank outputs.

At each step of the coding procedure $M = 32$ samples of the audio signal enter the M-band filter bank as shown in Fig. 10.5. These M samples together with the $L = 480$ previous samples form an input frame of length $N = L + M = 480 + 32 = 512$. Using these N samples the filter bank computes M transform coefficients. The output of the ith, $i = 0, 1, \ldots, M - 1$, filter is computed as

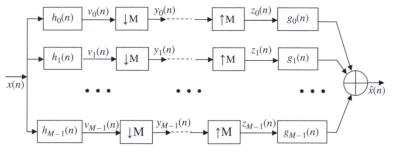

Figure 10.5 M-band analysis and synthesis filter banks

$$v_i(n) = \sum_{m=0}^{N-1} x(n-m)h_i(m) \tag{10.4}$$

where $x(n)$ are the input samples. After decimation by a factor M, we obtain

$$y_i(n) = v_i(Mn) = \sum_{m=0}^{N-1} x(Mn-m)h_i(m)$$

where $h_i(n)$ is the pulse response of the ith analysis filter. In the decoder the quantized transform coefficients $\hat{y}_i$ are upsampled by a factor M in order to form the intermediate sequences

$$z_i(n) = \begin{cases} \hat{y}_i(n/M), & n = 0, M, 2M, \dots \\ 0, & \text{otherwise.} \end{cases}$$

The obtained values $z_i(n)$ are then filtered by the synthesis filter bank where $g_i(n) = h_i(N-1-n)$, $i = 0, 1, \dots, M-1$ is the pulse response of the ith synthesis filter.

The ISO-MPEG coding uses a *cosine modulated* filter bank. The pulse responses of the analysis and synthesis filters are cosine modulated versions of the pulse response of a lowpass prototype filter; that is,

$$h_i(m) = h(m) \cos\left(\frac{\pi(i+1/2)(m-16)}{32}\right)$$

where

$$h(m) = \begin{cases} -w(m), & \text{if } \lfloor m/64 \rfloor \text{ is odd} \\ w(m), & \text{otherwise} \end{cases}$$

denotes the pulse response of the lowpass prototype filter, $w(m)$, $m = 0, 1, \dots, 511$, is the *transform window*. The pulse response $h(n)$ is multiplied by a cosine term to shift the lowpass pulse response $h(n)$ to the appropriate frequency band. These filters have normalized by f_s center frequencies $(2i+1)\pi/64$, $i = 0, 1, \dots, 31$, where f_s is the sampling frequency.

Although the presented form of the filter bank is rather convenient for analysis, it is not efficient from an implementation point of view. A direct implementation of (10.4) requires $32 \times 512 = 16\,384$ multiplications and $32 \times 511 = 16\,352$ additions to compute the 32 filter outputs. Taking into account the periodicity of the cosine function, we can obtain the equivalent, but computationally more efficient, representation

$$v_i(n) = \sum_{k=0}^{63} M(i, k) \sum_{j=0}^{7} w(k + 64j)x(n - (k + 64j)) \qquad (10.5)$$

where $w(k)$ is one of the 512 coefficients of the transform window,

$$M(i, k) = \cos\left(\frac{\pi(i + 1/2)(k - 16)}{32}\right)$$

$i = 0, 1, \ldots, 31, k = 0, 1, \ldots, 63$.

In order to compute the 32 filter outputs according to (10.5), we perform only $512 + 32 \times 64 = 2560$ multiplications and $64 \times 7 + 32 \times 63 = 2464$ additions or approximately 80 multiplications and additions per output.

The cosine modulated filter bank used by the MPEG standard does not correspond to any invertible transform. However, it can be shown that there exist generalized cosine modulated filter banks which provide perfect reconstruction of the input signal. Such filter banks can be implemented more efficiently by reducing the convolution to the MDCT which can be implemented using the FFT.

Layer III of the MPEG-1 standard uses the MDCT based on DCT-IV. The input of the transform block is blocks of samples $x_t = (x(n + tM), x(n + tM + 1), \ldots, x(n + tM + M - 1))$ of length M each. Together with the $(L - 1)$ previous blocks, they form a new block s_t of length $N = ML$. First, this block is componentwise multiplied by the analysis window $w_a(n), n = 0, 1, \ldots, N - 1$. Then, by multiplying the obtained vector v_t of length N by matrix A of size $N \times M$ (transform matrix), specified below, we obtain the vector y_t of M transform coefficients:

$$y_t = v_t A.$$

To perform the inverse transform, first, the vector of transform coefficients y_t of length M is multiplied by matrix A^{T}:

$$\hat{s}_t = y_t A^{\mathrm{T}}.$$

Then, the obtained vector $\hat{s}_t$ of length N is componentwise multiplied by the synthesis window $w_s(n), n = 0, 1, \ldots, N - 1$. The resulting vector $\hat{v}_t$ is shifted by M positions with respect to the previous block $\hat{v}_{t-1}$ and is added to the output sequence of the inverse transform block. Each block $\hat{x}$ at the output of the decoder is the sum of L shifted blocks of length N. The blocks are overlapped in $(L - 1)M$ samples. The overlapped analysis and overlap-add synthesis are shown in Fig. 10.6. The transform matrix A of size $N \times M$ can be written in the form

$$A = Y \cdot T$$

where matrix T of size $M \times M$ is the matrix of the DCT-IV transform (see Chapter 5 (5.30)) with entries

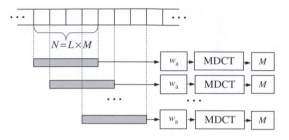

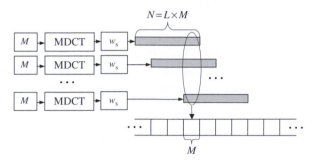

Figure 10.6   Modified DCT

$$t_{kn} = \sqrt{\frac{2}{M}} \cos\left(\left(k + \frac{1}{2}\right)\left(n + \frac{1}{2}\right)\frac{\pi}{M}\right)$$

$k, n = 0, 1, \ldots, M - 1$, matrix Y describes the preprocessing step. It has the following form

$$Y = [Y_0 \,|Y_1\,|-Y_0\,|-Y_1\,|Y_0\,|Y_1\, \ldots\,]^\mathrm{T}$$

where submatrices Y_0 and Y_1 of size $M \times M$ in turn consist of submatrices of size $M/2 \times M/2$:

$$Y_0 = \begin{pmatrix} 0 & 0 \\ I & -J \end{pmatrix}$$

$$Y_1 = \begin{pmatrix} -J & -I \\ 0 & 0 \end{pmatrix}$$

where I is the identity matrix of size $M/2 \times M/2$, the contra-identity matrix J is of the same size but has ones along the second diagonal and zeros elsewhere. To complete the description of the modified discrete cosine transform, it is necessary to specify the window functions $w_\mathrm{a}(n)$ and $w_\mathrm{s}(n)$. The analysis and synthesis windows can be chosen to be equal, $w_\mathrm{a}(n) = w_\mathrm{s}(n) = w(n)$. The requirement of the perfect reconstruction can be written as

$$\sum_{l=0}^{L-2s-1} w(n + lM)w(n + lM + 2sM) = \begin{cases} 1, s = 0 \\ 0, s \neq 0 \end{cases} \tag{10.6}$$

$n = 0, 1, \ldots, M/2 - 1$.

Often the MDCT with $N = 2M$ is used; that is, each block is overlapped with the adjacent block in half of the length. This corresponds to the parameter $L = 2$. In this case (10.6) takes the form

$$w^2(n) + w^2(n + M) = 1 \qquad (10.7)$$

$n = 0, 1, \ldots, M/2 - 1$.

The example of the window which satisfies (10.7) is

$$w(n) = \sin\left(\frac{\pi}{2N}\left(n + \frac{1}{2}\right)\right)$$

$n = 0, 1, \ldots, N - 1$.

All perceptual codecs based on transforms suffer from a specific artifact called *pre-echo*. This artifact occurs when, at the end of the transform block containing a low-energy region, a powerful signal with fast changes begins. Such a phenomenon is called *sound attack*. We compute the average SMR for the block. The decoder spreads the quantization error over the entire block. This quantization noise is easy to hear during the low-energy part of the signal. As a result we obtain in the reconstructed block an unmasked distortion in the low-energy region which precedes the sound attack in time. In order to suppress the pre-echo, the MPEG standard uses adaptive switching of the transform block length. The block length is larger for stationary parts of the signal and is reduced if the encoder detects a transient region.

Layer III of the MPEG standard improves performances of layers I and II by using the hybrid filter bank and adaptive switching of the transform block length for pre-echo suppression. The hybrid filter bank combines a 32-band polyphase filter bank with an adaptive MDCT. In order to improve the frequency resolution, each of the 32 subbands is subjected to an 18-point MDCT. Using this transform improves the frequency resolution from 750 Hz to 41.67 Hz. However, introducing the second step of transform increases the length of the transform block. To improve the pre-echo suppression, the 18-point MDCT is switched to a 6-point MDCT when a sound attack is detected.

For coding stereo signals, layers I and II of the MPEG standard use the *Intensity Stereo* method. The basic idea for intensity stereo is that for some high-frequency subbands, instead of separate transmitting signals of the left and right channels, only the sum of signals from these two channels is transmitted together with scale factors for each of the channels. Layer III supports intensity stereo coding but also uses a more efficient method of stereo coding called *Middle-Side* (*MS*)-stereo. The encoder uses the sum of the signals in the left and right channels (*middle information*) and the difference between them (*side information*).

The psychoacoustic model uses a separate, independent, time-to-frequency transform instead of the polyphase filter bank because it needs finer frequency resolution for accurate calculation of the masking thresholds. There are two psychoacoustic models in MPEG-audio; the so-called PAM-I and PAM-II. The PAM-I is less complex than the PAM-II and has more compromises to simplify the calculations. Each model works for any of the layers of compression. However, the PAM-I is recommended for layers I and

II while the PAM-II is recommended for layer III. The PAM-I uses a 512-point FFT for layer I and a 1024-point FFT for the layer II. PAM-II uses a 1024-point FFT for all layers.

The main steps of computing the signal-to-mask ratio are the following:

1. Calculate the sound pressure level (L_{sb}) for each subband as (Painter and Spanias 2000)

$$L_{sb} = \max_k \{P(k)\} \qquad (10.8)$$

where

$$P(k) = PN + 10\log_{10}\left(\sum_{n=0}^{N_F-1} w(n)x(n)e^{-j\frac{2\pi kn}{N_F}}\right)^2, 0 \le k \le \frac{N_F}{2}$$

$$w(k) = \frac{1}{2}\left(1 - \cos\left(\frac{2\pi k}{N_F}\right)\right)$$

is the Hann window, $PN = 90.302\,\text{dB}$ is the normalization term,

$$x(n) = \frac{s(n)}{N_F 2^{b-1}}$$

are normalized input audio samples $s(n)$, N_F is the FFT length, and b denotes the number of bits per sample. The maximum in (10.8) is computed over frequency samples in a given subband.

2. Calculate the absolute threshold of hearing $T_q(i)$ for each spectral component i of the given subband. Both models include an empirically determined absolute masking threshold; that is, the threshold in quiet.

3. Find tonal and noise-like spectral components.

Both models classify the spectral components of the audio signal into tonal and noiselike components, since the masking properties of these two types of signal differ. The spectral components are processed in groups corresponding to the critical bands. PAM-I identifies tonal components based on local peaks of the audio power spectrum. After processing the tonal components, PAM-I sums the remaining spectral values into a single nontonal component per critical band. The number i of this concentrated nontonal component is the value closest to the geometric mean of the corresponding critical band.

   Instead of classifying frequency components into tonal and nontonal, PAM-II computes a tonality index as a function of the frequency. This index is measured as a predictability of the component. It is known that tonal components are more predictable and thus have higher tonality indices. The value of the tonality index determines a degree in which the component is more tone-like or noise-like. PAM-II uses this index to interpolate between pure tone-masking-noise and noise-masking-tone values. The model uses data from the previous two analysis frames for linear prediction of the component values for the current frame.

4. Apply a spread function.

Both PAM-I and PAM-II take into account spreading the masking property of a given frequency component across its surrounding critical band. The models determine the noise-masking thresholds by first applying either an empirically determined masking function (PAM-I) or spreading function (PAM-II) to the signal components.

5. Compute the global masking threshold for each spectral component i,

$$T_g(i) = f\left(T_q(i), T_t(i, j), T_n(i, k)\right)$$

where $j = 1, 2, \ldots, m, k = 1, 2, \ldots, n$; $T_t(i, j), T_n(i, k)$ are masking thresholds for the tone and noise-like components, respectively, obtained by taking the spreading into account, m is the number of tonal thresholds, n is the number of noise-like thresholds, and f denotes an empirically found function.

6. Find the minimum masking threshold for a subband.

The PAM-I for each of the 32 subbands finds the minimum masking threshold,

$$T_{\min}(n) = \min_i \left\{T_g(i)\right\} \text{ dB} \tag{10.9}$$

where $n = 1, 2, \ldots, 32$, and the minimum is taken over all spectral components of the nth subband.

7. Calculate the SMR for each subband as

$$SMR(n) = L_{sb}(n) - T_{\min}(n). \tag{10.10}$$

Notice that steps 5 and 6 differ for PAM-I and PAM-II. According to PAM-II the minimum of the masking thresholds covered by the subband is chosen as a SMR only if the subband is wide compared to the critical band in that frequency region. PAM-II uses the average of the masking thresholds for the critical bands covered by the subband if the subband is narrow compared to the critical band.

For each subband the Mask-to-Noise Ratio (MNR) is then computed as follows,

$$MNR(n) = SNR(n) - SMR(n)$$

where $SMR(n)$ is obtained using PAM-I or PAM-II, and $SNR(n)$ is the SNR which is determined by the quantizer for the given subband. The standard describes the set of quantizers and the values of $SNR(n)$ are tabulated. An iterative procedure distributes the bits, starting with a given initial bit allocation until all $MNR(n)$ would be positive.

The purpose of MPEG-2 was to provide theater-style surround-sound capabilities. In addition to the right and left channels, there are a center channel in the front and left, and right channels in the rear or sides, denoted as the surround sound. There are actually five different operation modes corresponding to mono and two (left and right), three (left, right, and center), four (left, right, and rear surround), or five channels for the stereo audio signal.

A second goal of MPEG-2 was to achieve compatibility with MPEG-1. There are two types of compatibility. Forward compatibility means that MPEG-2 decoders can decode MPEG-1 bitstreams. Backward compatibility means that MPEG-1 decoders can decode a part of the MPEG-2 bitstream. This goal was achieved (1) by using MPEG-1 encoders and decoders as the component parts of MPEG-2, and (2) defining the MPEG-2 bitstream to consist of the MPEG-1 bitstream followed by a part of the bitstream which is only relevant to MPEG-2, i.e. coding the additional stereo channels. A matrix is used to mix the five component signals into five composite channels. The first two of these are coded by an MPEG-1 joint stereo coder. The other three are coded by the three-channel MPEG-2 extension coder. Full five-channel surround sound with MPEG-2 was evaluated to be transparent at a rate of 640 kb/s.

In 1997, the ISO-MPEG group developed a new Non Backwards-Compatible codec which was called MPEG-2-NBC. The quality of the synthesized five-channel stereo audio signals provided by this codec at a bit-rate of 320 kb/s was equivalent to the quality of the same signals provided by the MPEG-2 codec at a bit-rate of 640 kb/s. First, the new codec appeared as a part of the MPEG-2 standard; later, it was renamed the MPEG Advanced Audio Coder (MPEG-AAC).

When the MPEG-4 audio standard was introduced, AAC was included in this standard as two coding modes: HE-AAC also known as aacPlus technology and HE-AAC v2 (aacPlus v2 technology). These modes describe a very flexible audio codec with bit-rates in the range from 24 kb/s up to 256 kb/s. The sampling rates of its input signals vary from 8 kHz up to 96 kHz. It supports different mono and stereo coding modes. The AAC works on the basic principle of perceptual coding and has the same basic coding as the layer III of the MPEG-2. However, it has distinguishing features which significantly improve the quality of the synthesized audio signals compared to the MPEG-2 codecs.

Besides a filter bank with a higher frequency resolution and better entropy coding, it uses a new method of band expansion called *Spectral Band Replication* (SBR). This method exploits a correlation between the high- and low-frequency ranges of the original audio signal to approximate the high-frequency part by a properly scaled low-frequency part shifted to the high-frequency region. In other words, the high-frequency part of the signal is parameterized with respect to the low-frequency part of this signal. The corresponding parameters such as, for example, the spectral envelope of the original signal, called SBR data, are stored or transmitted as a part of the resulting bitstream. The HE-AAC v2 is a superset of HE-AAC-based codecs. It combines HE-AAC with an efficient method of coding stereo signals called *Parametric Stereo*. The basic idea of the Parametric Stereo is to parameterize the stereo signal in such a way that the bitstream will contain only compressed mono signal and parameters which enable a stereo signal to be synthesized from the reconstructed mono signal. Different types of parameter can be exploited; for example, inter-channel intensity difference, inter-channel correlation, inter-channel phase difference, etc.

MPEG-4 audio is a part of the MPEG-4 standard. It represents a new type of audio coding standard. It describes not a single or small set of highly efficient compression schemes but a complete toolbox such that it can provide low bit-rate speech coding as well as high-quality audio coding or music synthesis. The natural coding part within MPEG-4 audio describes traditional type speech and high-quality audio-coding algorithms, and their combination. The standard also includes new functionalities such as scalability.

There are several profiles containing natural audio object types:

- Speech coding profile. This profile contains the CELP, operating between 6 and 24 kb/s, with input speech sampled at 8 kHz and 16 kHz, respectively.
- Scalable audio profile is a superset of the Speech profile, suitable for scalable coding of speech and music and for different transmission methods, such as Internet and Digital Broadcasting. It contains the MPEG-2 AAC algorithm.
- The Main audio profile is a rich superset of the previous profiles (scalable, speech) containing tools for both natural and synthetic audio.

Problems

10.1   Transform a sequence of audio frames represented in the wav format by using a cosine modulated filter bank. Reconstruct the original sequence. Use frames of size 512.

10.2   Transform a sequence of audio frames represented in the wav format by using the MDCT. Reconstruct the original sequence. Use frames of size 1024.

10.3   Consider a frame of an audio signal containing $N = 1024$ samples. Transform the normalized audio samples by the FFT of length N. Split the obtained coefficients into 32 subbands. Compute the sound pressure level L_{sb} for each subband by (10.8). Compute the absolute threshold of hearing according to (10.1). Assume $T_g(i) = T_q(i)$, $i = 1, 2, \ldots, N$. Compute the minimum threshold for each subband by (10.9). Set to zero those frequency components of each subband which are less than the minimum threshold for the subband. Reconstruct the audio frame from the left frequency components. Compare the sound quality with the original audio frame.

10.4   Repeat Problem 10.3 for $N = 2048$. Compare the results.

10.5   (Project topic) Implement the following simplistic audio codec. For each audio frame of length $N = 1024$, 2048, or 4096 compute the MDCT spectrum coefficients $y = (y_1, y_2, \ldots, y_N)$. Split y into 32 subbands of equal width. Find the maximal value y^i_{max}, $i = 1, 2, \ldots, 32$ for each subband. Uniformly scalar quantize coefficients of each subband with quantization step δ_i which is computed as

$$\tilde{\delta}_i = \lceil y^i_{max}/q \rceil$$

where $q = 2$, 4, 8, or 16 and then changed by the closest power of 2. Let $m_i = \log_2 \delta_i$. Estimate the entropy of the quantized coefficients in each subband. Assuming that the sequence of m_i, $i = 1, 2, \ldots, 32$, is encoded by the unary or the Golomb code, estimate the compression ratio. Reconstruct the synthesized audio signal from the quantized coefficients. Compare the obtained quality/bit-rate tradeoff with your lovely audio codec.

10.6   (Project topic) Consider the following improvement of the codec from Problem 10.5: let the precision of the quantization q be dependent on the subband number. Take into account the minimum threshold for each subband. (For each audio frame of length

$N = 1024, 2048$, or 4096, compute the absolute threshold of hearing (10.1). Assume that $T_g(i) = T_q(i), i = 1, 2, \ldots, N$. Compute the minimum threshold for each subband by (10.9).) Repeat Problem 10.5. Compare the quality of the codecs.

10.7   (Project topic) Consider the following improvement of the codec from Problem 10.6: choose different subband widths (the subbands should be narrower for lower frequencies and wider for higher frequencies). Repeat Problem 10.6. Compare the quality of the codecs.

10.8   (Project topic) Consider the following improvement of the codec from Problem 10.7: use context coding for the quantized subband coefficients. Choose the sum of neighboring coefficients, with respect to the current coefficient, in each subband as a context. Repeat Problem 10.7. Compare the quality of the codecs.

A Lossless-coding techniques

In this Appendix we consider data compression algorithms which guarantee that the reconstructed file from the compressed bitstream will bit-by-bit coincide with the original input file. These *lossless* (entropy-coding) algorithms have many applications. They can be used for archiving different types of datum: texts, images, speech, and audio. Since multimedia data usually can be considered as outputs of a source with memory or, in other words, have significant redundancy, then the entropy coding can be combined with different types of preprocessing, for example, linear prediction.

However, multimedia compression standards are based mainly on lossy coding schemes which provide much larger compression ratios compared to lossless coding. It might seem that for such compression systems entropy coding plays no role but that is not the case. In fact, lossless coding is an important part of lossy compression standards also. In this case we consider quantized outputs of a preprocessing block as the outputs of a discrete-time source, estimate statistics of this source, and apply to its outputs entropy-coding techniques.

A.1 Symbol-by-symbol lossless coding

Let a random variable take on values x from the discrete set $X = \{0, 1, \ldots, M - 1\}$ and let $p(x) > 0$ be the probability mass function of X or, in other words, the probability distribution on the set X. If we do not take into account that different values x are not equally probable, we can only construct a fixed-length code for all possible values $x \in X$. The length of each codeword l coincides with the average codeword length l_{av} and is equal to $l_{av} = l = \log_2 M$ bits. However, we can reduce the average codeword length l_{av} if we will assign shorter codewords to more probable values x and longer codewords to less probable values x. It is known (Cover and Thomas 1971) that the lower bound on the average codeword length when using symbol-by-symbol variable-length coding is equal to the source entropy

$$H(X) = - \sum_{x=0}^{M-1} p(x) \log_2 p(x). \qquad (A.1)$$

The entropy $H(X)$ has the following properties:

- $H(X) \geq 0$.
- $H(X)$ is symmetric with respect to the distribution $p(0)$, $p(1)$, ..., $p(M-1)$, that is, it does not depend on the numbering of the values x.
- $H(X)$ achieves its maximum value $\log_2 M$ for the uniform distribution, i.e. $p(i) = 1/M$ for any $i \in \{0, 1, \ldots, M-1\}$.
- $H(X)$ achieves the minimum value 0 if $p(i) = 1$ for precisely one i.
- The entropy monotonically grows from 0 up to $\log_2(M)$ when α varies from 0 to $1/M$ where $p(i) = 1 - (M-1)\alpha$ for precisely one i and $p(j) = \alpha$ for all $j \neq i$.

Any method of symbol-by-symbol variable-length coding is characterized by the average codeword length

$$l_{av} = \sum_{x=0}^{M-1} p(x)l(x)$$

where $l(x)$ is the length of the codeword assigned to x. The difference $r = l_{av} - H(X)$ is called the code *redundancy*.

In order for a variable-length code to be uniquely decodable, its codewords usually satisfy special requirements, namely, none of the codewords is a *prefix*, that is, coincides with the beginning, of any other codeword. Notice that prefix or prefix-free codes are a subset of uniquely decodable codes, but in practice we deal only with prefix-free variable-length codes. There are a few approaches to constructing a prefix-free code. In 1952 (Huffman 1952) Huffman suggested an algorithm generating a prefix-free code that is optimum in the sense of minimizing the average codeword length l_{av}. The algorithm constructs a code tree using the following steps:

Input:
alphabet of size M
symbol probabilities
Output:
Huffman code

Initialization: Assign symbols to a set of M nodes; these nodes will be the leaf nodes of the code tree.

repeat

- Take the two nodes with the smallest probabilities and assign 1 and 0 arbitrarily to these two nodes.
- Merge these two nodes into a new auxiliary node whose probability is equal to the sum of the probabilities of the two merged nodes.

until *list of nodes contains one element*

Convert the code tree into a prefix-free code by tracing the assigned bits from the root of the tree to the leaves.

Algorithm A.1 Huffman code constructing algorithm

Table A.1 An example of the Shannon–Fano–Elias code

x	$p(x)$	$Q(x)$	$\sigma(x)$	Binary representation of $\sigma(x)$	$l(x)$	codeword
0	0.1	0.0	0.05	$0.00001\ldots$	5	00001
1	0.6	0.1	0.40	$0.01100\ldots$	2	01
2	0.3	0.7	0.85	$0.11011\ldots$	3	110

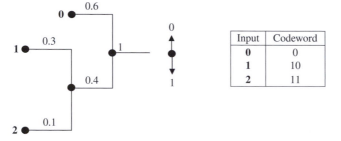

Figure A.1   An example of a Huffman tree code

A Huffman tree code for the probability distribution $\{p(0) = 0.6,\ p(1) = 0.3,$ $p(2) = 0.1\}$ is shown in Fig. A.1. The average codeword length is $l_{av} = 1.4$ bits, the entropy $H(X) = 1.3$ bits, and the code redundancy $r = 1.4 - 1.3 = 0.1$ bits. It is proved (Gallager 1978) that in general the redundancy of Huffman coding can be upper-bounded as $r \le p_{max} + 0.086$ where p_{max} denotes the probability of the most probable source symbol.

Another method of symbol-by-symbol variable-length coding known as Shannon–Fano–Elias coding (Cover and Thomas 1971) can be described as follows. We associate with each symbol x the modified cumulative function

$$\sigma(x) = Q(x) + \frac{p(x)}{2}$$

where

$$Q(x) = \sum_{a \prec x} p(a) \tag{A.2}$$

is the cumulative sum associated to x and $a \prec x$ denotes a symbol preceding x in the lexicographic order. It is easy to see that, according to (A.2), $Q(0) = 0$, $Q(1) = p(1)$, and $Q(M-1) = \sum_{a=0}^{M-2} p(a)$. The $\lfloor \sigma(x) \rfloor_{l(x)}$ is a codeword for x where $l(x) = -\lceil \log_2 p(x)/2 \rceil$ and $\lfloor N \rfloor_l$ denotes rounding off N to l bits. An example of the Shannon–Fano–Elias code constructed for the probability distribution $\{p(0) = 0.1,\ p(1) = 0.6,$ $p(2) = 0.3\}$ is given in Table A.1.

A graphical interpretation of Shannon–Fano–Elias coding is given in Fig. A.2. For details, see, for example, Kudryashov (2009). Assume that $x = 1$ is transmitted. The coding procedure maps the input value into a point inside the interval $(0, 1)$. The first

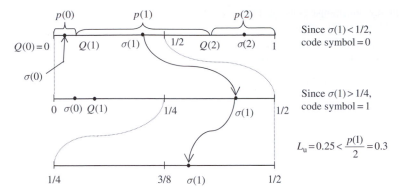

Since $\sigma(1) < 1/2$,
code symbol $= 0$

Since $\sigma(1) > 1/4$,
code symbol $= 1$

$$L_u = 0.25 < \frac{p(1)}{2} = 0.3$$

Figure A.2 Graphical interpretation of Shannon–Fano–Elias coding

symbol of the constructed codeword specifies to which half of the interval (0, 1) the value $\sigma(1)$ belongs. Since $\sigma(1) < 0.5$, the first code symbol is 0. The second symbol of the codeword specifies to which half of the interval (0, 0.5) belongs $\sigma(1)$. Since $\sigma(2) > 0.25$, the second code symbol is 1. At each step of the coding procedure the length of the interval of uncertainty L_u around $\sigma(1)$ is equal to 2^{-s} where s is the number of binary symbols that have been already transmitted. We stop the procedure if L_u is less than or equal to $p(x)/2$ since this condition guarantees that we will be able uniquely to decode x. In our example we stop after two steps, since $L_u = 0.25 < p(1)/2 = 0.3$. We obtain $l_{av} = 2.6$, that is, $r = 2.6 - 1.3 = 1.3$ bits. In general, the redundancy of Shannon–Fano–Elias coding satisfies the following inequalities $1 \leq r \leq 2$.

We have considered two symbol-by-symbol coding procedures. Huffman coding is optimal in the sense of coding redundancy. Shannon–Fano–Elias has significantly larger redundancy than Huffman coding. The following question arises: why do we need a nonoptimal coding procedure with larger redundancy? The answer is related to the computational complexity of the considered procedures. It is easy to see that the Shannon–Fano–Elias coding procedure has lower computational complexity than the Huffman coding procedure. In order to obtain the codeword for a given input symbol, this procedure does not construct the whole code tree as it is supposed to be done by Huffman coding. Moreover, in the next section we will show that the difference in the computational complexities drastically increases when we generalize the considered coding procedures to block coding.

A.2 Lossless block coding

Now we will try to explain in which cases symbol-by-symbol coding is not efficient. Assume that we deal with a memoryless binary source with the probability distribution $\{p(0) = 0.99, p(1) = 0.01\}$. The entropy of such a source is $H(X) = 0.081$. However, applying Huffman coding to this source, we obtain $l_{av} = 1$; that is, $r = 0.919$. In general, if the entropy $H(X) << 1$ the redundancy of symbol-by-symbol coding is rather

Table A.2 An example of the Huffman code for blocks of length 2

Input sequence	$p(x)$	Codeword
00	0.36	1
01	0.18	011
02	0.06	0010
10	0.18	010
11	0.09	0001
12	0.03	00001
20	0.06	0011
21	0.03	000001
22	0.01	000000

large, since we cannot have codewords of length less than 1 and the average length l_{av} is always greater than or equal to 1.

In order to reduce coding redundancy, we should use block lossless coding. Let $x \in X = \{0, 1, \ldots, M - 1\}$. We encode blocks $x = (x_1, x_2, \ldots, x_n)$ which appear at the output of the source during n consecutive time moments. Each block x of length n can be interpreted as an enlarged output symbol of a new source with the alphabet X^n where X^n is the set of all possible M^n vectors of length n with components from the set X. For a memoryless source $p(x) = \prod_{i=1}^{n} p(x_i)$. We can apply any symbol-by-symbol coding procedure to the output blocks (enlarged symbols) of the new source. It is known that the lower bound on the average code rate $R_{av} = l_{av}/n$ is equal to $H(X^n)/n$ where

$$H(X^n) = -\sum_{x \in X^n} p(x) \log_2 p(x).$$

Let r_n be the redundancy of the chosen block coding procedure, then we have

$$l_{av} = H(X^n) + r_n \tag{A.3}$$

$$R_{av} = \frac{H(X^n) + r_n}{n} = \frac{H(X^n)}{n} + \frac{r_n}{n}. \tag{A.4}$$

By taking into account that for a memoryless source $H(X^n) = nH(X)$ we obtain

$$R_{av} = H(X) + \frac{r_n}{n}. \tag{A.5}$$

It is evident that $r_n \leq r$ where r is the redundancy of the corresponding method for symbol-by-symbol coding. Therefore, $r_n \leq c$ where c is a constant depending on the method of symbol-by-symbol coding. From (A.5) we obtain that if the block length n tends to infinity, the per symbol redundancy of block coding r_n/n tends to zero.

An example of the Huffman code constructed for blocks of length 2 at the output of the memoryless source with the probability distribution $\{p(0) = 0.6, \ p(1) = 0.3, \ p(2) = 0.1\}$ is presented in Table A.2. For this example, $H(X^2) = 2.60$ bits, $l_{av} = 2.68$ bits, $r_2 = 0.08$ bits, $R_{av} = 1.34$ bits, and $H(X) = 1.30$ bits.

For stationary discrete sources with memory $H(X^n) \leq n H(X)$ and therefore $H(X^n)/n \leq H(X)$. From (A.4) we can conclude that the coding gain of block coding compared to symbol-by-symbol coding for the stationary sources with memory is even larger than for memoryless sources. Notice that

$$\lim_{n \to \infty} \frac{H(X^n)}{n} = H_\infty(X) \tag{A.6}$$

where $H_\infty(X)$ is called the *entropy rate* of the stationary discrete source with memory.

Now we should answer the question: is it possible to implement block coding with acceptable computational complexity? Unfortunately, for Huffman block coding the answer is negative. The alphabet of the new source is of size M^n. It means that the coding complexity of the Huffman tree code will grow exponentially with the block length. In the next section we will describe a block coding procedure which is a direct extension of Shannon–Fano–Elias coding. This procedure is called *arithmetic coding* and can be implemented recurrently with complexity proportional to n^2.

A.3   Arithmetic coding

Arithmetic coding is a generalization of the Shannon–Fano–Elias coding scheme. Let $\boldsymbol{x} = (x_1, x_2, \ldots, x_n)$ be an M-ary sequence of length n. We associate with $\boldsymbol{x}$ the modified cumulative distribution function

$$\sigma(\boldsymbol{x}) = \sum_{\boldsymbol{a} \prec \boldsymbol{x}} p(\boldsymbol{a}) + \frac{p(\boldsymbol{x})}{2} = Q(\boldsymbol{x}) + \frac{p(\boldsymbol{x})}{2}$$

where $\boldsymbol{a} \prec \boldsymbol{x}$ means that the sequence $\boldsymbol{a}$ is lexicographically less than $\boldsymbol{x}$. Consider two sequences $\boldsymbol{x} = (x_1, x_2, \ldots, x_n)$ and $\boldsymbol{y} = (y_1, y_2, \ldots, y_n)$ from X^n. We say that $\boldsymbol{x} \prec \boldsymbol{y}$ if $y_i \prec x_i$ and i is the smallest index such that $x_i \neq y_i$. The codeword for $\boldsymbol{x}$ is obtained as $\lfloor \sigma(\boldsymbol{x}) \rfloor_{l(\boldsymbol{x})}$, where $l(\boldsymbol{x}) = -\lceil \log_2 p(\boldsymbol{x})/2 \rceil$.

Assume that we deal with a memoryless source. In this case

$$p(\boldsymbol{x}) = \prod_{i=1}^{n} p(x_i). \tag{A.7}$$

Equation (A.7) can be easily rewritten in the recurrent form

$$p(\boldsymbol{x}_1^n) = p(\boldsymbol{x}_1^{n-1}) p(x_n) \tag{A.8}$$

where $\boldsymbol{x}_1^i = (x_1, x_2, \ldots, x_i)$.

Now we derive a recurrent formula for computing

$$Q(\boldsymbol{x}_1^n) = \sum_{\boldsymbol{a}_1^n \prec \boldsymbol{x}_1^n} p(\boldsymbol{x}_1^n).$$

We split the sum into two sums: the first sum includes such sequences a_1^n which contain such subsequences a_1^{n-1} which are lexicographically less than x_1^{n-1}, the second sum includes such sequences a_1^n which contain subsequences $a_1^{n-1} = x_1^{n-1}$ and the symbol $a_n \prec x_n$. By taking into account (A.7) we obtain the following chain of inequalities

$$Q(x_1^n) = \sum_{a_1^{n-1} \prec x_1^{n-1}} \sum_{a_n} p(a_1^{n-1} a_n) + \sum_{a_1^{n-1} = x_1^{n-1}} \sum_{a_n \prec x_n} p(a_1^{n-1} a_n)$$

$$= \sum_{a_1^{n-1} \prec x_1^{n-1}} p(a_1^{n-1}) + \sum_{a_1^{n-1} = x_1^{n-1}} p(a_1^{n-1}) \sum_{a_n \prec x_n} p(a_n)$$

$$= Q(x_1^{n-1}) + p(x_1^{n-1})Q(x_n) \tag{A.9}$$

where $Q(x_n)$ is the cumulative sum associated with x_n. Using the obtained recurrent formulas we can formulate the following coding procedure:

Input:
alphabet size M
symbol probabilities $p(0), p(1), ..., p(M-1)$
input sequence $(x_1, x_2, ..., x_n)$
Output:
codeword encoding $(x_1, x_2, ..., x_n)$

Initialization:
Set $F = 0$, $G = 1$, $Q(0) = 0$;
compute the cumulative sums
for $j = 1$ **to** $M - 1$ **do**
 | $Q(j) = Q(j-1) + p(j-1)$
end

1. **for** $i = 1$ **to** n **do**
   | $F \leftarrow F + Q(x_i) \times G$;
   | $G \leftarrow G \times p(x_i)$;
   end
2. $F \leftarrow F + \frac{G}{2}$, $l = -\lceil \log_2 G/2 \rceil$
3. Construct the codeword $c \leftarrow \lfloor F \rfloor_l$.

Algorithm A.2   Arithmetic coding

Assume that the probability distribution is $\{p(0) = 0.1, p(1) = 0.6, p(2) = 0.3\}$ and the input sequence of length $n = 5$ is **12101**. The calculations performed by the arithmetic encoder are shown in Table A.3. The graphical interpretation of arithmetic coding is illustrated in Fig. A.3.

Table A.3 Implementation of arithmetic coding

Step i	x_i	$p(x_i)$	$Q(x_i)$	F	G
0	–	–	–	0.0000	1.0000
1	1	0.6	0.1	0.1000	0.6000
2	2	0.3	0.7	0.5200	0.1800
3	1	0.6	0.1	0.5380	0.1080
4	0	0.1	0.0	0.5380	0.0108
5	1	0.6	0.1	0.5391	0.0065
	Length of the codeword $= \lceil -\log G + 1 \rceil = 9$				
6	Codeword $= F + G/2 = 0.5423... \rightarrow$				
	$\lfloor F \rfloor_l = 0.541 \rightarrow 100010101$				

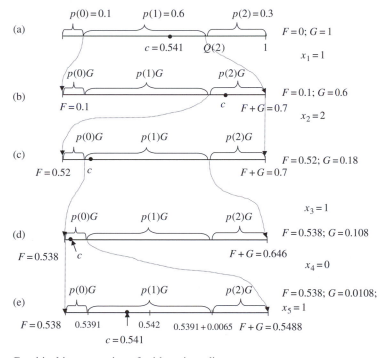

Figure A.3   Graphical interpretation of arithmetic coding

At each step of the coding procedure we recalculate the beginning F and the length G of the interval which contains the number c corresponding to the codeword for the input sequence **12101**. Since $x_1 = 1$ we obtain that c belongs to the interval $(Q(1), Q(1) + p(1)) = (0.1, 0.7)$. This interval is shown in Fig. A.3(b). Since $x_2 = 2$, then the beginning of the interval is moved to $Q(1) + p(1)Q(2) = 0.52$ and the length of the interval is reduced to $G = p(1)p(2) = 0.18$, etc. After the fifth step $F = 0.5391$. By adding the bias $G/2$ and rounding off the result to the length $l = 9$ bits, we obtain $c = 0.541$. Thus, the input sequence is mapped into one point c of the interval $(0, 1)$. It can be shown that the closest possible point corresponding to another input sequence is located at least $1/2^9 = 1/512$ away from c. It guarantees that the input sequence will be uniquely decoded from the obtained codeword.

The computational complexity of the coding procedure can be estimated as follows. At each step of the coding procedure we perform one addition and two multiplications. Moreover, at each step the complexity of each operation increases with increasing required length of the binary representation of the operands. Assume that the probabilities $p(0)$, $p(1)$, ..., $p(M - 1)$ are represented by b-bit values. After the first step, F and G have to be represented by $2b$-bit values. After n steps the maximal length the binary representation can achieve is nb bits. Thus, the total complexity is

$$b + 2b + \cdots + nb = \frac{n(n + 1)b}{2} \approx bn^2,$$

i.e. grows with n as n^2.

A.4   Decoding issues

We will start with decoding of the Shannon–Fano–Elias code. Let $Q(0)$, $Q(1)$, ..., $Q(M - 1)$ be cumulative sums of the alphabet symbols and $\hat{\sigma} = \lfloor \sigma \rfloor_l$ be the input of the decoder.

The decoding algorithm is presented below:

Input: $\hat{\sigma}$
Output: decoded symbol x

Initialization: Set $Q(M) = 1, m = 0$;
while $Q(m + 1) < \hat{\sigma}$ **do**
  $|$ $m \leftarrow m + 1$;
end
Output is $x = m$

Algorithm A.3   Decoding of the Shannon–Fano–Elias code

Starting with $m = 1$ at each step of the decoding procedure, we compare the cumulative sum $Q(m)$ with the decoder input $\hat{\sigma}$. We stop the procedure if the cumulative sum is larger than $\hat{\sigma}$ and output $x = m - 1$. Similarly, the decoder of the arithmetic code recurrently computes the value of $Q(x)$ closest to the decoder input $\hat{F}$. The output of the decoder is the corresponding sequence x. The decoding algorithm is presented below:

Input:
alphabet of size M
symbol probabilities $p(0), p(1), \ldots, p(M-1)$
number $\hat{F}$ representing encoded data
Output:
decoded data $\boldsymbol{x} = (x_1, x_2, \ldots, x_n)$

Initialization:
Set $Q(M) = 1, S = 0, G = 1$;
for $i = 1$ **to** n **do**
  $j = 0$;
  while $S + Q(j+1)G < \hat{F}$ **do**
   $j \leftarrow j + 1$
  end

  • $S \leftarrow S + Q(j)G$;
  • $G \leftarrow p(j)G$;
  • $x_i = j$

end
Output is $\boldsymbol{x} = (x_1, x_2, \ldots, x_n)$

Algorithm A.4 Arithmetic code decoding

The calculations performed by the decoder of the arithmetic code for the input sequence 100010101 are presented in Table A.4.

Although the computational complexity of arithmetic coding is proportional to the squared length of the input sequence, the procedure looks, at first glance, completely impractical. The main problems related to the implementation of arithmetic coding are:

• Theoretically arithmetic coding requires infinite computational precision to ensure that the input sequence is accurately encoded.
• The coding delay is equal to the length of the input sequence.

It transpired that both problems can be solved and arithmetic coding can be implemented in 32-bit and even 16-bit integer arithmetics. The detailed description of the corresponding algorithm and C-program of the arithmetic coder and decoder can be found in (Witten *et al.* 1987). Here we will consider only the main ideas behind the approach proposed in (Witten *et al.* 1987). As shown above, the arithmetic encoder is trying to specify inside the interval $(0, 1)$ the position of the point corresponding to the input sequence. At each step of the coding procedure the length of the interval of uncertainty is reduced. The values $Low = F$ and $High = F + G$ are the lower and upper bounds of the current interval, respectively. These lower and upper bounds will slowly converge to the output value. If the most significant bits of Low and $High$ coincide, then these most significant bits can be output by the encoder. For example, assume that at some step of the coding procedure $Low = 0$ and $High = 0.001100000$, that is, both the lower and upper bounds are less than 0.5. It means that the output value will

Table A.4 Decoding of the sequence from example in Table A.3

Step	S	G	Hypo-thesis x	$Q(x)$	$S + QG$	Solu-tion x_i	$p(x)$
0			$100010101 \rightarrow \hat{F} = 0.541$				
			0	0.0	$0.0000 < \hat{F}$		
1	0.0000	1.0000	**1**	0.1	$\mathbf{0.1000} < \hat{F}$	**1**	0.6
			2	0.7	$0.7000 > \hat{F}$		
			0	0.0	$0.1000 < \hat{F}$		
2	0.1000	0.6000	**1**	0.1	$0.1600 < \hat{F}$	**2**	0.3
			2	0.7	$\mathbf{0.5200} > \hat{F}$		
			0	0.0	$0.5200 < \hat{F}$		
1	0.5200	0.1800	**1**	0.1	$\mathbf{0.5380} < \hat{F}$	**1**	0.6
			2	0.7	$0.6460 > \hat{F}$		
1	0.5380	1.0800	**0**	0.0	$\mathbf{0.5380} < \hat{F}$	**0**	0.1
			1	0.1			
			0	0.0	$0.5380 < \hat{F}$		
1	0.5200	0.0108	**1**	0.1	$\mathbf{0.5391} < \hat{F}$	**1**	0.6
			2	0.7	$0.5456 > \hat{F}$		

never be larger than 0.5. We transmit 0 and normalize the lower and upper bounds as follows:

$$Low \leftarrow Low \times 2 = 0$$

$$High \leftarrow High \times 2 = 0.011000000.$$

Since both bounds are still less than 0.5 we repeat the same steps: transmit 0 and normalize Low and $High$. Now assume that $Low = 0.110000010$ and $High = 0.111100000$. It is clear that in this case the output value will never be less than 0.5. We transmit 1 and normalize Low and $High$ as follows:

$$Low \leftarrow (Low - 0.5) \times 2 = 0.100000100$$

$$High \leftarrow (High - 0.5) \times 2 = 0.111000000. \tag{A.10}$$

Since both bounds are still larger than 0.5 we repeat the same steps: transmit 1 and normalize Low and $High$ according to (A.10). The problem appears when Low is close to 0.5, but less than 0.5, for example, $Low = 0.011111\ldots$ and $High > 0.5$. Since the most significant bits of Low and $High$ are not equal, we cannot output the corresponding bit and normalize Low and $High$. In this case the encoder continues with the calculations and increments the counter until either a 0 bit will appear

($Low = 0.01111 \ldots 10$) and $High$ will become less than 0.5 or Low will become greater than 0.5; that is, we will have one of the two previously considered situations. Depending on whether Low and $High$ converged to the lower or higher value, the encoder outputs $0111 \ldots 1$ or $1000 \ldots 0$ where the run length is equal to the value accumulated by the counter.

A.5 Context lossless coding for sources with memory

We already mentioned that speech, image, audio, and video signals can be considered as outputs of sources with memory. Even the quantized transform coefficients are dependent and can be interpreted as symbols of a discrete source with memory. The simplest model of a discrete source with memory is the so-called *Markov chain* of a given order, s say. The probability of a random vector $\mathbf{X} = (X_1, X_2, \ldots, X_n), n = 1, 2, \ldots,$ for such a discrete source is determined as follows:

$$p(\mathbf{x}) = p(x_1, x_2, \ldots, x_s) p(x_{s+1} | x_1, x_2, \ldots, x_s)$$

$$\times p(x_{s+2} | x_2, x_3, \ldots x_{s+1}) \ldots p(x_n | x_{n-s}, x_{n-s+1}, \ldots, x_{n-1})$$

or, in other words, we can say that for $n > s$

$$p(x_n | x_1, x_2, \ldots, x_{n-1}) = p(x_n | x_{n-s}, x_{n-s+1}, \ldots, x_{n-1}).$$

For stationary Markov chains the probabilities $p(\mathbf{x})$ do not depend on a shift in time. The conditional probabilities also do not depend on a shift in time, that is,

$$p(x_n = i_1 | x_{n-1} = i_2, x_{n-2} = i_3, \ldots, x_{n-s} = i_{s+1})$$

$$= p(x_{n+j} = i_1 | x_{n+j-1} = i_2, x_{n+j-2} = i_3, \ldots, x_{n+j-s} = i_{s+1})$$

where j is a shift by j symbols.

Taking into account the property of stationarity, we introduce the following notation

$$H(X|X^n) = H(X_{n+1} | X_1, X_2, \ldots, X_n) \tag{A.11}$$

where $H(X|Y) = - \sum_{x \in X} \sum_{y \in Y} p(x, y) \log_2 p(x|y)$ denotes the conditional entropy of X given Y. It is easy to see that in the left-hand side we simply omitted time indices, since for a stationary source its n-dimensional conditional entropy does not depend on a shift in time and X^n here denotes a block of the n successive source outputs preceding the given symbol X.

The entropy rate of a stationary discrete source with memory is defined by (A.6). Another definition for the entropy rate of the stationary discrete source with memory (Cover and Thomas 1971) is

$$H(X|X^\infty) = \lim_{n \to \infty} H(X|X_n). \tag{A.12}$$

For stationary sources both limits (A.6) and (A.12) exist and coincide.

It can be shown (Cover and Thomas 1971) that for a stationary Markov chain of order s the conditional entropy can be determined as follows:

$$H(X|X^n) = H(X_{n+1}|X_1, X_2, \ldots, X_n) =$$

$$= H(X_{n+1}|X_{n-s+1}, X_{n-s}, \ldots, X_n) = H(X|X^s)$$

and the entropy rate for such a source is

$$H_\infty(X) = H(X|X^s) \leq H(X)$$

where $H(X)$ is the one-dimensional entropy of the Markov chain determined by (A.1). Typically, $H(X|X^s)$ is significantly less than $H(X)$ so in order to compress multimedia data efficiently it is important to approach $H(X|X^s)$ by lossless coding procedures. In principle, it can be done by using, for example, arithmetic coding. We should modify formulas (A.7), (A.8), and (A.9) as follows:

$$p\left(\boldsymbol{x}_1^n\right) = p\left(\boldsymbol{x}_1^{n-1}\right) p\left(x_n|\boldsymbol{x}_{n-s}^{n-1}\right)$$

$$Q\left(\boldsymbol{x}_1^n\right) = Q\left(\boldsymbol{x}_1^{n-1}\right) + p\left(\boldsymbol{x}_1^n\right) Q\left(x_n|\boldsymbol{x}_{n-s}^{n-1}\right)$$

where

$$Q\left(x_n|\boldsymbol{x}_{n-s}^{n-1}\right) = \sum_{a \prec x_n} p\left(a|\boldsymbol{x}_{n-s}^{n-1}\right).$$

However, in order to use such arithmetic coding, for each symbol of the source alphabet we need to know q^s conditional probabilities where q is the alphabet size. It means that we must either transmit a lot of auxiliary information about the conditional probability distribution or somehow estimate this distribution. Surely, this circumstance reduces the efficiency of arithmetic coding and significantly increases its computational complexity. A practical solution of this problem is to use so-called *context coding*.

Context coding implies that the source output data are split into substreams according to the context values. The context value is known for the decoder since the context depends on the already received symbols. Each of the obtained substreams can be approximately considered as the output of a memoryless source. In general, the answer to the question: "How should we choose the context?" is unknown. For example, consider a stationary Markov chain of order s. Assume that

$$\boldsymbol{x}_1^{t-1} = (x_1, x_2, \ldots, x_{t-1})$$

are already transmitted and it is necessary to transmit

$$\boldsymbol{x}_t^n = (x_t, x_{t+1}, \ldots, x_n).$$

Since for a Markov chain of order s we have

$$p(x_t|x_1, x_2, \ldots, x_{t-1}) = p(x_t|x_{t-s}, x_{t-s+1}, \ldots, x_{t-1})$$

and we can choose $(x_{t-s}, x_{t-s+1}, \ldots, x_{t-1})$ as a context C_t for the tth symbol. It is not necessary that C_t takes on one of q^s values $\{c_1, c_2, \ldots, c_{q^s}\}$ since if $p(x_t|c_i) \approx p(x_t|c_j)$ we join the corresponding context values. Assume that $C_t \in \{c_1^t, c_2^t, \ldots, c_K^t\}$ where $K \leq q^s$, then we can write

$$\sum_{i=1}^{K} p\left(c_i^t\right) H\left(X_t|c_i^t\right) \leq H(X)$$

since conditioning always reduces entropy. On the other hand, we have

$$H\left(X_t|C_t\right) \geq H\left(X_t|X_1, X_2, \ldots, X_{t-1}\right)$$

$$= H(X|X^s),$$

i.e. typically context coding has a redundancy compared to the theoretical limits.

A.6   QM-coder

We showed in the previous sections that arithmetic coding is a very efficient lossless-coding technique. It can even be applied to the sources with memory without drastic loss in coding efficiency. However, the main problem related to this method is its relatively high computational complexity. We have already considered some ideas which allow us to implement arithmetic coding in fixed-precision arithmetic. Further simplification of the coding and decoding procedures can be obtained by using binary arithmetic coding. Actually, all modern lossy multimedia compression standards use binary arithmetic coding. Moreover, these binary arithmetic codecs are usually implemented as multiplication-free approximate arithmetic codecs. The binary arithmetic coder of this type is called the *QM-coder* (Pennebaker and Mitchell 1993). We will consider it in this section.

The QM-coder classifies each input binary symbol as More Probable Symbol (MPS) and Less Probable Symbol (LPS).   In order to perform such a classification the QM-coder uses a context. If the predicted, according to the context, bit value does not coincide with the input bit value, then this bit is classified as LPS, otherwise, it is classified as MPS. As was shown before, arithmetic coding is an iterative procedure of finding the position of the point corresponding to the input sequence in the interval $(0,1)$. We initialize the lower bound of the interval (Low) by 0 and the upper bound ($High$) by 1. Then at each step of the procedure we recompute Low and $High$ according to the input sequence. For implementation in fixed-precision arithmetic instead of the interval $(0,1)$ we use the normalized interval $(0, maxint)$, where $maxint = 2^b - 1$ and b is a length of the binary representation in a given arithmetic. Now consider the specific features of the binary arithmetic coder which allow further simplifications of the implementation.

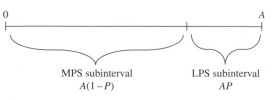

Figure A.4   Division of the probability interval

Denote the probability of the LPS by P, then the probability of the MPS is $1 - P$. The cumulative probabilities of the MPS and LPS are $Q(\text{MPS}) = 0$ and $Q(\text{LPS}) = 1 - P$, respectively. We set $Low = 0$ and $High = A$ where A is the length of the interval. Figure A.4 shows how the probability interval A is split into the MPS subinterval $(0, A(1 - P))$ and LPS subinterval $(A(1 - P), A)$.

If the next input symbol is MPS, then $Low \leftarrow Low + 0$, that is, it is simply kept unchanged and the length of the interval is recomputed as $A \leftarrow A \times (1 - P)$, otherwise $Low \leftarrow Low + A \times (1 - P)$ and $A \leftarrow A \times P$.

It is easy to see that at each step of the procedure the length of the interval A becomes smaller. It was explained before that, in order to keep the fixed precision of the computations, the value A can be renormalized by doubling each time it becomes too small. The doubling is performed as a left shift and does not require a multiplication operation. However, to modify A at each step of the procedure we need the multiplication operation. This operation is rather time and computation-consuming.

In order to speed up the computations the QM-coder uses an approximation of the multiplication operation. The idea is to keep the value A close to $1(maxint)$. If this is the case the operation $A \leftarrow A \times (1 - P)$ is approximated by $A \leftarrow A - P$ and $A \leftarrow A \times P$ is approximated by $A \leftarrow P$. The requirement that A should be close to 1 determines the threshold that we use for normalization. Notice that if A will be doubled each time it gets a little below 1, then after doubling, $2A$ can be closer to 2 than to 1. Moreover, we should choose the threshold, taking into account the fixed precision of the used arithmetic. In the QM-coder the threshold for normalization is chosen to be 0.75.

One more point which is important in the implementation of the QM-coder is the so-called *interval inversion*. This problem arises from the approximation of the multiplication operation. As a result, the subinterval allocated to the MPS sometimes becomes less than the interval allocated to the LPS. It can happen when P is close to 0.5 and $0.75 < A < 1$. In this case A is not normalized and if the next symbol is MPS, then the length of the MPS subinterval is updated as $A - P < P$. This problem is solved by an exchange of the subintervals called the *conditional exchange*. Notice that the condition for the interval inversion occurs when $A - P < P < 0.5$; that is, both the MPS and LPS subintervals are less than 0.75. It means that renormalization must be performed. This circumstance is taken into account in the QM-coder and the interval inversion condition is checked after it is found that the renormalization is needed. Thus, the coding procedure can be described as follows:

Table A.5 QM encoding

Symbol	*Low*	*A*	*A* normalized	*Low* normalized
–	0	1	1	0
LPS	$0 + 1 - 0.2 = 0.8$	0.2	0.8	$0.8 \cdot 2^2 = 3.2$
MPS	3.2	$0.8 - 0.2 = 0.6$	1.2	$3.2 \cdot 2 = 6.4$
LPS	$6.4 + 1.2 - 0.2 = 7.4$	0.2	0.8	$7.4 \cdot 2^2 = 29.6$
MPS	29.6	$0.8 - 0.2 = 0.6$	1.2	$29.6 \cdot 2 = 59.2$

Input: input binary sequence
Output: encoded sequence

while *not end of data to be encoded* **do**
    if *MPS* **then**
        Low is unchanged
        $A \leftarrow A - P$
        if $A < 0.75$ **then**
            if $A < P$ **then**
                $Low \leftarrow Low + A$
                $A \leftarrow P$
            end
            renormalize A, Low
        end
    else
        $A \leftarrow A - P$
        if $A \geq P$ **then**
            $Low \leftarrow Low + A$
            $A \leftarrow P$
        end
        renormalize A, Low
    end
end
generate encoded sequence from *Low*

Algorithm A.5   QM-coding procedure

In Table A.5 the results of applying this coding procedure to the sequence of four bits are presented. The probability P of the LPS is equal to 0.2.

The decoding procedure for the same sequence is demonstrated in Table A.6. At each step of the decoding procedure we compare Low with the threshold $A(1 - P) \approx A - P$. If $Low \leq A - P$, then the decision is MPS, otherwise LPS. After MPS we do not change Low and do renormalize A as $A - P$. After LPS we change Low by $Low - A + P$ and renormalize A as AP.

Table A.6 QM decoding

Threshold $A - P$	Decision	*Low*	A	A normalized	*Low* normalized
–	–	59.2	1	1	$59.2 \cdot 2^{-6} = 0.9250$
$0.8 \leq Low$	LPS	$0.925 - 1 + 0.2 = 0.125$	0.2	0.8	$0.125 \cdot 2^2 = 0.5$
$0.6 > Low$	MPS	0.5	0.6	1.2	$0.5 \cdot 2 = 1.0$
$1.0 \leq Low$	LPS	$1 - 1.2 + 0.2 = 0.0$	0.2	0.8	0.0
$0.6 > Low$	MPS	0.0	0.6	1.2	0.0

The QM-coder also uses an efficient approach to estimate the probability P of the LPS. It is based on a table of precomputed P values. However, we do not consider this approach here.

A.7   Monotonic codes

Lossless coding methods such as Huffman coding and arithmetic coding require knowledge of the probability distribution on the source symbols. Thus, we should either transmit estimates of the symbol probabilities or somehow adapt "on-the-fly" the probability estimates used by the coder and decoder. In this section we consider so-called "monotonic codes" which do not use any knowledge of the probability distribution and can be implemented with very low computational complexity. They are often used in video compression standards. Codes of this class are prefix-free codes which map the positive integers onto binary codewords with the additional property that for any $i, j \in \{1, 2, \ldots\}$ from $i < j$ follows $l_i \leq l_j$ where l_i is the codelength for i.

A.7.1   Unary code

Denote by $s^m, s \in \{0, 1\}$ a run of m symbols s. The unary code maps i into the codeword $0^{i-1}1$. The codewords of the unary code for 1, 2, 3 are $unar(1) = 1, unar(2) = 01$, and $unar(3) = 001$, respectively. For the number i the length of codeword l_i is equal to i.

It can be shown that this code is optimal on the one-sided geometric probability distribution; that is,

$$p_i = (1 - \alpha)\alpha^{i-1}, i = 1, 2, \ldots$$

when the parameter $\alpha = 1/2$. If $\alpha > 1/2$, then the *Golomb code* (Golomb 1966) is better than the unary code.

A.7.2   Golomb code

Let us introduce a parameter $T = 2^m$ and assume that the number $i \in \{1, 2, \ldots\}$ is mapped into a codeword that consists of two parts: prefix and suffix. The prefix is equal to *unar* ($\lfloor i/T \rfloor + 1$). The suffix is the binary representation of $i \mod T$ of length m. The length of the codeword for the number i is $l_i = \lfloor i/T \rfloor + 1 + m$. For example, let m be equal to 4 and i be equal to 23, then the prefix is $unar(\lfloor 23/2^4 \rfloor + 1) = unar(2)$ and the suffix is $23 \mod 2^4 = 7$, that is, the codeword is 010111. In literature the Golomb codes with the parameter $T = 2^m$ are also called the Golomb–Rice codes.

A.7.3   Gallager–vanVoorhis codes

This code (Gallager and VanVoorhis 1975) is a generalization of the Golomb code where T is not equal to a power of 2. The number i is mapped into the codeword that also consists of a prefix and suffix. The prefix is equal to *unar* ($\lfloor i/T \rfloor + 1$), as for the Golomb code. The suffix is computed as $i \mod T$ and coded by the variable-length code with codewords of lengths $m = \lfloor \log_2 T \rfloor$ and $m + 1$. For example, if $T = 6$, then the suffix values are 0, 1, 2, 3, 4, 5 which are coded as 00, 01, 100, 101, 110, and 111. In other words, we assume that all of the values 0, 1, 2, 3, 4, 5 are equiprobable and we encode them by the Huffman code. For $i = 23$, we obtain the prefix equal to *unar* ($\lfloor 23/6 \rfloor + 1$) = *unar*(4), the suffix is equal to $23 \mod 6 = 5$ which is encoded as 111. Thus, 23 is encoded by the codeword 0001111.

The optimal value of T for the Gallager–vanVoorhis codes is determined by

$$\alpha^{T+1} + \alpha^T \le 1 < \alpha^{T-1} + \alpha^T \tag{A.13}$$

where α is a parameter of geometric probability distribution.

The Gallager–vanVoorhis codes were initially introduced for coding run lengths. It is easy to see that for the discrete source of independent binary symbols, the probabilities of runs (1, 01, ..., $0^l 1$, ...,) satisfy a one-sided geometric distribution. It was shown in (Gallager and VanVoorhis 1975) that the Golomb codes are optimal for the discrete binary sources with one-sided geometric distribution. In other words, for such sources with infinite alphabet, the Golomb codes provide the minimum average codeword length over all symbol-by-symbol lossless coding procedures.

Choosing the parameter T according to (A.13) is equivalent to the approximate condition

$$\alpha^T \approx \frac{1}{2}.$$

It means that we choose T in such a way that the probability of a zero run of length T is approximately equal to 1/2. The codeword of the Golomb code with parameter T for 0^T is 0, i.e. has length one!

When using the Golomb codes for nonbinary discrete sources with one-sided geometric distribution, i.e. $p(x) = (1 - \alpha)\alpha^x$, $x = 0, 1, 2, \ldots$, we consider runs (1, 01, ..., $0^x 1$, ...,) as binary representations of $x = 0, 1, 2, \ldots$ For α close to 1 we can rewrite (A.13) as

$$T \approx -\frac{\ln 2}{\ln \alpha} = -\frac{\ln 2}{\ln (1 - (1 - \alpha))} \approx \frac{1}{(1 - \alpha)}. \tag{A.14}$$

Taking into account that $1/(1 - \alpha)$ is the mathematical expectation of the geometric distribution, we obtain that the parameter T in this case should be approximately equal to the mathematical expectation of the distribution, i.e.

$$T \approx \mathrm{E}\{x\}.$$

A.7.4   Levenstein codes

These codes were introduced by (Levenstein 1965). We will explain the idea behind the Levenstein code by an example. Let i be equal to 23. The binary representation of i is 10111. We cannot use 10111 directly as a codeword for i, since in this case the obtained code will not be a prefix-free code. In order to construct a prefix code, we add to the binary representation of i a prefix equal to the length of this binary representation encoded by the unary code. In our example the length of the sequence 10111 is equal to 5. Thus, we obtain the codeword $(00001)(10111) = 0000110111$. In general, the length of the codeword of this code will be equal to $2\lceil \log_2 i \rceil$. We can improve this code by taking into account that the first bit of the binary representation of any number i is 1. We can omit this 1 and obtain the codeword of monotonic code in the form $(00001)(0111) = 000010111$. We can further reduce the length of the codewords if we represent the length of the binary representation in binary form without the first one, then add the length of this representation in the binary form and so on. We continue iterations until one digit is left. In order to obtain a prefix-free code, we then add to the codeword the number of iterations plus 1 coded by the unary code. For $i = 23$ we obtain

$$unar(4)(0)(00)(0111) = (0001)(0)(00)(0111) = 00010000111$$

since the number of iterations is equal to 3. The length of the codeword of the Levenstein code is upperbounded by

$$\log_2 i + 2\log_2(1 + \log_2 i) + 2.$$

The decoder first decodes the unary code and finds that there were three iterations. Then it reads 0 and after adding prefix 1 obtains 10. It decodes 10 and finds that the number of bits in the binary representation of the length of the codeword is equal to 2. Then it reads 00 and adding prefix 1 obtains 100. It means that the length of the binary representation is equal to 4. Finally, it reads 0111, adds prefix 1 and decodes 10111, that is, obtains 23.

The Elias code (Elias 1987) is a simplified version of the Levenstein code. Any codeword of the Elias code consists of three parts. The first part from the right-hand side part is the binary representation of the corresponding number without the first one, the second is the binary representation of the length of the first part without the first one, and the last is the codeword of the unary code for the length of the second part increased by 2. For example, for $i = 18$, we obtain:

$$(0001)(00)(0010) = 0001000010.$$

A.8 Binarization

Binarization is a one-to-one mapping of a nonbinary symbol into a binary sequence. These binary sequences are chosen to form a prefix-free code. The high implementation efficiency of the binary arithmetic coder and its multiplication-free versions motivates having the binarization step before the VLC in modern multimedia compression standards (Marpe *et al.* 2003). Moreover, binarization allows context modeling on a subsymbol level. The following four binarization types are used, for example, in the H.264 standard:

- unary binarization: an unsigned integer is coded by the unary code, i.e. $x \geq 0$ is encoded as $unar(x + 1)$;
- truncated unary binarization: it is defined only for $0 \leq x \leq M$; if $x = M$, then the terminating "1" is omitted.
- kth-order Golomb binarization: an unsigned integer is coded by the Golomb code with a given parameter k, i.e. $T = 2^k$;
- fixed-length binarization: an unsigned integer $0 \leq x < M$ is given by its binary representation of length $\lceil \log_2 M \rceil$ bits.

The most critical issue related to the binarization step is the assignment of probabilities to the bits of the obtained binary sequence. In order to avoid any efficiency loss of the VLC that follows the binarization, these probabilities are calculated as explained below. Binarization can be interpreted as a preliminary coding of an input nonbinary symbol by a prefix-free code. Any prefix-free code can be represented in the form of a binary tree with leaves corresponding to possible values of the input nonbinary symbol. Each path from the root to a leaf represents a codeword for the given symbol value. The corresponding trees for the unary and fixed-length binarization schemes are shown in Figs A.5 (a) and (b), respectively.

To guarantee the same redundancy of the VLC, the nonbinary symbol probabilities have to be uniquely reconstructed from the bit probabilities. An example of a binarization tree with assigned bit probabilities is shown in Fig. A.5(b). Each node of the tree is labeled by the unconditional probability of a part of a bitstream corresponding to the path leading to the node. Each branch of the path is labeled by a conditional bit probability. It is easy to see that the unconditional probabilities of the leaves coincide with the probabilities of the corresponding nonbinary symbols. For example,

$$P(4) = P(100) = P(1)P(0|1)P(0|10).$$

Thus, the binary variable-length encoder should be provided by the corresponding conditional bit probabilities. Another example of the binarization tree for the unary binarization is shown in Fig. A.5(a). The probability of symbol 4 corresponding to a leaf of this tree is calculated as

$$P(4) = P(00001) = P(0)P(0|0)P(0|00)P(0|000)P(1|0000).$$

The most computationally efficient binarization can be performed by using Huffman codes. Such a binarization tree minimizes the average codeword length and thereby the

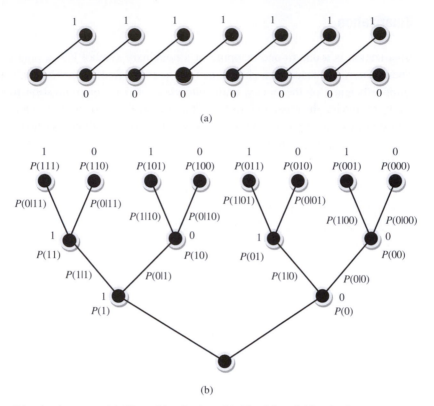

Figure A.5   Binarization trees: (a). Unary binarization, (b). Fixed-length binarization

average number of computations is reduced. However, taking into account the necessity to adapt the binarization scheme to the varying statistics of the source, the simpler binarization schemes listed earlier are often used in practice. They allow an online computation of binary representations and do not need storing of any tables.

A.9   Adaptive coding

The previously considered lossless coding algorithms such as Huffman coding, Shannon–Fano coding, and arithmetic coding work with a given probability distribution. However, in practice, we typically do not know this distribution. Moreover, we often deal with nonstationary sources, that is, with sources whose probability distribution can vary in time. Thus, one more important problem which we should solve to implement these lossless coding procedures is how to assign probabilities to the source symbols.

There are a few solutions of this problem for the stationary sources. Assume that x_1, $x_2, \ldots, x_t$ is a sequence observed at the encoder input. We estimate the probability that the symbol x_{t+1} is equal to a given value a using the preceding sequence $x_1, x_2, \ldots, x_t$. At first glance it seems that the probability of the symbol x_{t+1} can be estimated as

$$\hat{P}(x_{t+1} = a) = \frac{N_t(a)}{t} \tag{A.15}$$

where $N_t(a)$ is the number of occurrences of the value a in the sequence of length t. However, applying (A.15) to estimating symbol probabilities we can sometimes obtain zero estimates. Existence of zero probabilities in the source probability distribution is not critical for Huffman coding but it is not acceptable for Shannon–Fano or arithmetic coding. For this reason different biased symbol probability estimates are used. For example, we can estimate the probability $P(x_{t+1} = a)$ as follows

$$\frac{N_t(a) + 1}{t + Q} \tag{A.16}$$

where Q denotes the alphabet size of the source. In (A.16) we avoid zero probabilities by adding one to the numerator. In order to preserve the sum of probability estimates equal to one, we add the alphabet size to the denominator. For large t this estimate is rather close to (A.15). However, for short source sequences (A.16) gives estimates which are significantly less than those obtained by (A.15). This leads to using longer codewords and thereby reduces the efficiency of lossless coding.

Another approach to estimating the probability distribution consists in introducing to the alphabet an additional symbol called *escape symbol*. If at the $(t + 1)$th step the symbol x_{t+1} is equal to one of the alphabet symbols which already occurred among the preceding t symbols, then the following probability is assigned to x_{t+1}

$$\hat{P}(x_{t+1} = a) = \frac{N_t(a)}{t + 1}.$$

If at the $(t + 1)$th step a new alphabet symbol occurred, then we transmit the escape symbol followed by this symbol. To the escape symbol we assign the probability

$$\hat{P}_{t+1}(\text{esc}) = \frac{1}{t + 1}$$

and to the symbol x_{t+1} the probability

$$\frac{1}{Q - Q_t}$$

is assigned, where Q_t denotes the number of different alphabet symbols which already occurred in the sequence $x_1, x_2, \ldots, x_t$. Thus, we use the following estimate

$$\hat{P}(x_{t+1} = a) = \begin{cases} \frac{N_t(a)}{t+1} & \text{if } N_t(a) > 0 \\ \frac{1}{t+1} \frac{1}{Q-Q_t} & \text{if } N_t(a) = 0. \end{cases}$$

For nonstationary sources, probability estimates are usually computed by using not all preceding symbols but only preceding symbols in a rather short window.

In some codecs the probability estimation is implemented as a prediction algorithm. Such a prediction can be based on a finite-state machine. Depending on the previous state and an input symbol, the prediction algorithm determines the next state that represents a probability estimate corresponding to the next input symbol. For example, such a finite-state machine with 112 states is used by the QM-coder in the JBIG coding standard (see Chapter 8). The states of the finite-state machine are switched depending on the input symbol value. The probability estimates associated with different states as well as the state transitions are tabulated. Similar ideas are used by CABAC exploited by the H.264 standard (see Chapter 8). The finite-state machine for each context model in CABAC can be in one of 128 states.

References

M. Antonini, M. Barlaud, P. Mathieu, and I. Daubechies. Image coding using wavelet transform. *IEEE Trans. Image Processing*, **1**(2): 205–220, 1992.

L. R. Bahl, J. Cocke, F. Jelinek, and J. Raviv. Optimal decoding of linear codes for minimizing symbol error rate. *IEEE Trans. Infn. Theory*, **20**(2): 284–287, 1974.

T. Berger. *Rate Distortion Theory: A Mathematical Basis for Data Compression*. Prentice-Hall, Inc., Englewood Cliffs, NJ, 1971.

R. E. Blahut. Computation of channel capacity and rate-distortion functions. *IEEE Trans. Infn. Theory*, **18**(4): 460–473, 1972.

I. Bocharova, V. Kolesnik, B. Kudryashov, *et al.* Two-dimensional hierarchical quantizing and coding for wavelet image compression. In *Proceedings of the Eighth International Conference on Signal Processing, Applications and Technology, ICSPAT*, pages 1233–1237, San-Diego, CA, USA, 14–17, Sept., 1997.

I. Bocharova, V. Kolesnik, B. Kudryashov, *et al.* Memory-saving wavelet based image/video compression algorithm. In *Proceedings of the 10th International Conference on Signal Processing, Applications and Technology, ICSPAT*, Orlando, FL, USA, 4–7, Nov., 1999.

A. R. Calderbank and N. J. Sloane. New trellis codes based on lattices and cosets. *IEEE Trans. Infn. Theory*, **33**(2): 177–195, 1987.

R. V. Churchill and J. W. Brown. *Complex Variables and Applications*. McGraw-Hill, NY, 1990.

A. Cohen, I. Daubechies, and J. C. Feauveau. Biorthogonal bases of compactly supported wavelets. *Comm. Pure&Appl. Math.*, **45**(5): 485–560, 1992.

J. H. Conway and N. J. A. Sloane. *Sphere Packings, Lattices and Groups*. Springer-Verlag, NJ, 1988.

T. M. Cover. Enumerative source coding. *IEEE Trans. Infn. Theory*, **19**(1): 73–77, 1973.

T. M. Cover and J. A. Thomas. *Elements of Information Theory*. Wiley & Sons, Inc., NY, 1991.

R. E. Crochiere and L. R. Rabiner. *Multirate Digital Signal Processing*. Prentice-Hall, Inc., Upper Saddle River, NJ, 1983.

I. Csiszar and G. Tusnady. Information geometry and alternating minimization procedures. *Statists and Decisions. Suppl. Issue*, **1**: 205–237, 1984.

I. Daubechies. Orthogonal bases of compactly supported wavelets. *Comm. Pure&Appl. Math.*, **41**(7): 909–996, 1988.

J. Durbin. The fitting of time series models. *Rev. Inst. Int. Statists*, **28**: 233–243, 1960.

P. Elias. Interval and recency rank source coding: two on-line adaptive variable-length schemes. *IEEE Trans. Infn. Theory*, **33**(1): 3–10, 1987.

N. Farvardin and J. W. Modestino. Optimum quantizer performance for a class of non-gaussian memoryless sources. *IEEE Trans. Infn. Theory*, **30**(3): 485–497, 1984.

G. D. Forney. Convolutional codes II: maximum likelihood decoding. *Infn. Control*, **25**(3): 222–266, 1974.

CCITT recommendation G.711. Pulse code modulation (PCM) of voice frequencies, 1972.

ITU-T recommendation G.723.1. Dual rate coder for multimedia telecommunications transmitting at 5.3 and 6.4 Kbit/s, 1993.

CCITT recommendation G.726. 40, 32, 24, 16 Kbit/s adaptive differential pulse code modulation, 1990.

R. Gallager. Variations on a theme by Huffman. *IEEE Trans. Infn. Theory*, **24**(6): 668–673, 1978.

R. G. Gallager and D. C. VanVoorhis. Optimal source codes for geometrically distributed integer alphabets. *IEEE Trans. Infn. Theory*, **21**(2): 228–230, 1975.

A. Gersho and R. M. Gray. *Vector Quantization and Signal Compression*. Kluwer Academic Publishers, Norwell, MA, USA, 1992.

H. Gish and J. N. Pierce. Asymptotically efficient quantizing. *IEEE Trans. Infn. Theory*, **14**(5): 676–683, 1968.

S. W. Golomb. Run-length encodings. *IEEE Trans. Infn. Theory*, **12**(3): 399–401, 1966.

ITU-T Recommendation H.261. Video CODEC for audiovisual services at $p \times 64$ Kbit/s, 1993.

ITU-T Recommendation H.263. Video codec for low bit rate communication, 2005.

ITU-T Recommendation H.264. Advanced video coding for generic audiovisual services, 2007.

J. J. Y. Huang and P. M. Schultheiss. Block quantization of correlated Gaussian random variables. *IEEE Trans. Comm.*, **11**(3): 289–296, 1963.

D. A. Huffman. A method for the construction of minimum redundancy codes. *Proc. IRE*, **40**(9): 1098–1101, 1952.

N. S. Jayant and P. Noll. *Digital Coding of Waveforms: Principles and Applications to Speech and Video*. Prentice Hall Inc., NJ, 1984.

ISO/IEC International Standard 10918-1 JPEG. Information technology – digital compression and coding of continuous-tone still images: requirements and guidelines, 1994.

JPEG-2000. ISO/IEC International Standard 15444. Information technology – JPEG 2000 image coding system, 2000.

V. N. Koshelev. Quantization with minimal entropy. *Probl. Pered. Infn.*, **14**: 151–156, 1963 (in Russian).

V. A. Kotelnikov. *On the Carrying Capacity of the Ether and Wire in Telecommunications*. Izd. Red. Upr. Svyazi RKKA, Moscow, 1933 (in Russian).

H. P. Kramer and M. V. Mathews. A linear coding for transmitting a set of correlated signals. *IEEE Trans. Infn. Theory*, **2**(3): 41–46, 1956.

B. D. Kudryashov. *Information Theory*. Piter, St Petersburg, 2009 (in Russian).

B. D. Kudryashov and A. V. Porov. Scalar quantizers for random variables with the generalized gaussian distribution. *Digital Signal Processing*, (4): 2005 (in Russian).

B. D. Kudryashov and K. V. Yurkov. Random coding bound for the second moment of multidimensional lattices. *Probl. Infn Transm.*, **43**(1): 57–68, 2007.

V. I. Levenstein. On redundancy and delay of the separable coding of natural numbers. *Problemy Kibernetiki*, (14): 173–179, 1965 (in Russian).

N. Levinson. The Wiener RMS error criterion in filter design and prediction. *J. Math. Phys.*, 25: 261–278, 1947.

S. Lin and D. Costello, Jr. *Error Control Coding: Fundamentals and Applications*. Prentice Hall, Upper Saddle River, NJ, 2nd edn, 2004.

Y. Linde, A. Buzo, and R. Gray. An algorithm for vector quantizer design. *IEEE Trans. Commun.*, **28**(1): 84–94, 1980.

S. G. Mallat. A theory for multiresolution signal decomposition: the wavelet representation. *IEEE Trans. Pattern Analysis and Mach. Intell.*, **7**(11): 674–693, 1989.

S. G. Mallat. *Wavelet Tour of Signal Processing*. Academic Press, 2nd edn, 1999.

M. W. Marcellin and T. R. Fischer. Trellis-coded quantization of memoryless and Gauss-Markov sources. *IEEE Trans. Commun.*, **38**(1): 82–93, 1990.

D. Marpe, H. Schwarz, and T. Wiegand. Context-based adaptive binary arithmetic coding in the H.264/AVC video compression standard. *IEEE Trans. Circuits and Systems for Video Technol.*, **13**(7): 620–636, 2003.

J. Max. Quantizing for minimum distortion. *IEEE Trans. Infn. Theory*, **6**(1): 7–12, 1960.

S. K. Mitra. *Digital Signal Processing: Computer-based Approach*. McGraw-Hill, NY, 3rd edn, 2005.

ISO/IEC International Standard 11172 MPEG-1. Information technology – coding of moving pictures and associated audio for digital storage media at up to about 1.5 Mbit/s, 1993.

ISO/IEC International Standard DIS 13818 MPEG-2. Generic coding of moving pictures and associated audio information, 1999.

ISO/IECJTC1/SC29/WG11 N 4668 International Standard MPEG-4. Coding of moving pictures and audio, 2002.

ISO/IEC 14496-10 MPEG4-AVC. Information technology – coding of audio-visual objects – Part 10: advanced video coding, 2005.

P. Noll and R. Zelinski. Bounds on quantizer performance in the low bit-rate region. *IEEE Trans. Commun.*, **26**(2): 300–304, 1978.

H. Nyquist. Certain topics in telegraph transmission theory. *Trans. AIEE*, **47**: 617–644, 1928.

B. Oliver, J. Pierce, and C. Shannon. The philosophy of P. C. M. *Proc. IRE*, **36**(11): 1324–1331, 1948.

T. Painter and A. Spanias. Perceptual coding of digital audio. *Proc. IEEE*, **88**(4): 451–515, 2000.

W. B. Pennebaker and J. L. Mitchell. *JPEG: Still Image Data Compression Standard*. Springer-Verlag, NY, 1993.

L. R. Rabiner and R. W. Schafer. *Digital Processing of Speech Signals*. Prentice-Hall, Englewood Cliffs, NJ, 1978.

I. E. G. Richardson. *H.264 and MPEG-4 Video Compression*. Wiley, 2003.

M. S. Roden. *Communication Systems Design*. Prentice-Hall, Englewood Cliffs, NJ, 1988.

A. Said and W. A. Pearlman. A new, fast, and efficient image codec based on set partitioning in hierarchical trees. *IEEE Trans. Circuits and Systems for Video Technol.*, **6**(3): 243–250, 1996.

J. M. Shapiro. Embedded image coding using zerotrees of wavelet coefficients. *IEEE Trans. Signal Processing*, **41**(12): 3445–3462, 1993.

G. Strang and T. Nguyen. *Wavelets and Filter Banks*. Wellesley–Cambridge Press, Wellesley, MA, 1996.

G. Ungerboeck. Channel coding with multilevel/phase signals. *IEEE Trans. Infn. Theory*, **28**(1): 55–67, 1982.

A. J. Viterbi. Error bounds for convolutional codes and an asymptotically optimum decoding algorithm. *IEEE Trans. Infn. Theory*, **13**(2): 260–269, 1967.

A. J. Viterbi and J. K. Omura. *Principles of Digital Communications and Coding*. McGraw-Hill, Inc., NY, 1979.

M. J. Weinberger, G. Seroussi, and G. Sapiro. LOCO-I: a low complexity, context-based, lossless image compression algorithm. In *Proc. DCC'96*, pp. 140–149, Snowbird, Utah, USA, Mar. 1996.

E. T. Whittaker. On the functions which are represented by the expansions of the interpolation theory. *Proc. Royal Soc., Edinburgh*, Sec. **A**, 35: 181–194, 1915.

I. H. Witten, R. M. Neal, and J. G. Cleary. Arithmetic coding for data compression. *Commun. ACM*, **30**(6): 530–540, 1987.

P. Zador. Asymptotic quantization error of continuous signals and their quantization dimension. *IEEE Trans. Infn. Theory*, **28**(2): 139–149, 1982.

Index